ENVIRONMENTAL SCIENCE, ENGINEERING AND TECHNOLOGY SERIES

FOREST CANOPIES: FOREST PRODUCTION, ECOSYSTEM HEALTH AND CLIMATE CONDITIONS

ENVIRONMENTAL SCIENCE, ENGINEERING AND TECHNOLOGY SERIES

The Amazon Gold Rush and Environmental Mercury Contamination
Daniel Marcos Bonotto and Ene Glória da Silveira
2009. ISBN: 978-1-60741-609-8

Environmental Effects of Off-Highway Vehicles
*Douglas S. Ouren, Christopher Haas, Cynthia P. Melcher, Susan C. Stewart,
Phadrea D. Ponds, Natalie R. Sexton, Lucy Burris,
Tammy Fancher and Zachary H. Bowen*
2009. ISBN: 978-1-60692-936-0

Nitrous Oxide Emissions Research Progress
Adam I. Sheldon and Edward P. Barnhart (Editors)
2009. ISBN: 978-1-60692-267-5

Agricultural Runoff, Coastal Engineering and Flooding
Christopher A. Hudspeth and Timothy E. Reeve (Editors)
2009. ISBN: 978-1-60741-097-3

**Fundamentals and Applications of Biosorption Isotherms,
Kinetics and Thermodynamics**
Yu Liu and Jianlong Wang (Editors)
2009. ISBN: 978-1-60741-169-7

Conservation of Natural Resources
Nikolas J. Kudrow (Editor)
2009. ISBN: 978-1-60741-178-9

Forest Canopies: Forest Production, Ecosystem Health and Climate Conditions
Jason D. Creighton and Paul J. Roney (Editor)
2009. ISBN: 978-1-60741-457-5

ENVIRONMENTAL SCIENCE, ENGINEERING AND TECHNOLOGY SERIES

FOREST CANOPIES: FOREST PRODUCTION, ECOSYSTEM HEALTH AND CLIMATE CONDITIONS

JASON D. CREIGHTON

AND

PAUL J. RONEY

EDITORS

Nova Science Publishers, Inc.

New York

LIBRARY OF CONGRESS CATALOGING-IN-PUBLICATION DATA
Forest canopies : forest production, ecosystem health, and climate conditions / [edited by] Jason D. Creighton and Paul J. Roney.
    p. cm.
 Includes bibliographical references and index.
 ISBN 978-1-60741-457-5 (hardcover : alk. paper)
 1. Forest canopies. 2. Forest canopy ecology. I. Creighton, Jason D. II. Roney, Paul J.
 QH541.5.F6F662   2009
 577.3--dc22
                                                                                      2009010237

Published by Nova Science Publishers, Inc. ✦ New York

CONTENTS

Preface   vii

Chapter 1   A New Paradigm of Forest Canopy Interception Science: The
                 Implication of a Huge Amount of Evaporation During Rainfall   1
                 Shigeki Murakami

Chapter 2   Exotic Herb Layers as Ecological Filters in Forest Understories   29
                 Christopher R. Webster, Michael A. Jenkins,
                 Shibu Jose and Linda M. Nagel

Chapter 3   Quantitative Analysis of Canopy Photosynthesis
                 Influenced by Light Simulation Models   51
                 Toru Sakai, Hiroyuki Muraoka, Tsuyoshi Akiyama,
                 Michio Shibayama and Yoshio Awaya

Chapter 4   Lidar Remote Sensing for Forest Canopy Studies   71
                 A. Farid, D.C. Goodrich and S. Sorooshian

Chapter 5   Soil Organic Carbon Dynamics of Different
                 Land Use in Southeast Asia   85
                 Minaco Adachi and Hiroshi Koizumi

Chapter 6   Carbon Stable Isotopes of Mammal Bones as Tracers of Canopy
                 Development and Habitat Use in Temperate and Boreal Contexts   103
                 Dorothée G. Drucker and Hervé Bocherens

Chapter 7   Simulating the Two-Way Feedback between
                 Terrestrial Ecosystems and Climate: Importance of
                 Forest Ecological Processes on Global Change   111
                 Takeshi Ise, Tomohiro Hajima, Hisashi Sato and Tomomichi Kato

Chapter 8   Atmospheric Deposition and its Leaf Surface
                 Interactions in Japanese Cedar Forests   127
                 Hiroyuki Sase and Takejiro Takamatsu

Chapter 9   Effects of Forest Canopy Gaps on Litter Microarthropod
                 Populations in the Southern Appalachians   143
                 Cynthia C. Kaminski, Steve Patch, and Barbara C. Reynolds

Chapter 10   Interactions Between Urban Vegetated
                 Surfaces and the Atmosphere   **153**
                 Timo Vesala, Leena Järvi, Üllar Rannik,
                 Sampo Smolander, Andrey Sogachev, and Eero Nikinmaa

Index   **161**

PREFACE

Forests cover approximately 30% of total land area and function as habitats for organisms, hydrologic flow modulators, and soil conservers, constituting one of the most important aspects of the Earth's biosphere. The canopy is one of the uppermost levels of a forest, below the emergent layer, formed by the tree crowns. The canopy is home to unique flora and fauna not found in other layers of a forest. Trees in the canopy are able to photosynthesise very rapidly thanks to the large amount of light, so it supports the widest diversity of plant as well as animal life in most rainforests. This book presents a wide variety of topics on the ecosystem in forest canopies. Included is a study on light distribution patterns and how it effects the daily photosynthesis of herbaceous vegetation. Recent progress, concerns, and future directions in simulations of vegetation processes are presented as well, in the terrestrial biosphere model that is coupled to a climate system model.

Chapter 1 - Canopy interception includes some enigmatic problems. Firstly, evaporation rate is proportional to the rainfall intensity. Secondly, the heat budget model underestimates the measured canopy interception at some sites but the Gash models do not, though both types of models are physically based. Thirdly, though high evaporation rate during rainfall, e.g. ≥ 10 mm h^{-1}, was observed, the source of latent heat of vaporization and efficient water vapor transfer mechanism during rainfall remain unsolved. The first problem, the dependence of canopy interception on rainfall intensity, DOCIORI, cannot be explained by the conventional heat budget model, since rainfall intensity is not included as a parameter. To elucidate the phenomenon, the idea of splash droplet evaporation, SDE, was proposed. As the raindrop size and the number of raindrops per unit volume increase with the rainfall intensity, so does the number of splash droplets produced by the raindrops hitting on the canopy. Splash droplets evaporate efficiently due to large combined surface area. The DOCIORI was pointed out in the Japanese sites for the first time in the 1980s, but has not been known worldwide until recently. In this chapter it is shown that the DOCIORI occurs at sites other than in Japan using literature data in which authors did not notice the phenomenon. The second problem is also understandable by SDE; the heat budget model considers evaporation from canopy surface only but actual evaporation occurs from the surface of splash droplets as well. This overlook leads to underestimation of the heat budget model especially where rainfall intensity is sufficiently high, while estimation by the Gash models reproduces the measurement fairly well, because they include SDE implicitly. The third problem concerns not only forest hydrology but also atmospheric sciences, and is explainable at least qualitatively by the innovative idea of evaporative force proposed recently, which is an unnoticed basic principle in meteorology.

Chapter 2 - A growing body of evidence suggests that exotic herb layers in forest understories are fundamentally altering ecological processes and successional dynamics. The establishment of mono-dominant patches of exotic plants is often influenced by complex interactions between propagule pressure, disturbance, climate, and land-use legacies. We use a combination of field studies and an extensive literature review to explore invasion ecology and consequences of invasion for the perpetuation of native forests along a gradient extending from southern Florida to the northern Lake States. Based on the commonalities between these invasions, we propose a general framework for integrating invasive species detection and control into forest management activities.

Chapter 3 - Light distribution pattern is a critical input for canopy photosynthesis models. We examined how and to what extent the different calculations of light distribution affect estimations of daily photosynthesis (A_{day}) of herbaceous vegetation, *Sasa senanensis*, in a deciduous forest understory. A_{day} was estimated using three models, the M-S$_1$ (the Monsi-Saeki model 1; diurnal incident light was defined by sine curve from sunrise to sunset and the vertical profile of incident light through the stand was calculated using the Beer's law), M-S$_2$ (a model combining the hemispherical photographic method to describe sunfleck/diffuse light above the stands and Beer's law to describe the vertical profile of incident light within the stands) and Y-plant (incident light was calculated for every single leaves by using a hemispherical photograph and geographical distribution of single leaf area and orientation). Although estimations of daily light absorption by the whole stands were relatively close among three models, A_{day} varied greatly in order of the M-S$_1$, M-S$_2$ and Y-plant models under clear sky conditions. The significant overestimation of A_{day} was attributed to the different manner of calculating (i) the diurnal pattern of incident light above the stands and (ii) the light distribution pattern within the stands. The use of diurnal pattern of incident light estimated by a sine curve led to a 76% overestimation of A_{day} in the model separating sunflecks/diffuse light (i.e., the M-S$_1$ model vs the M-S$_2$ model). In addition, less calculation of the clumping effect such as leaf overlaps caused a 78% overestimation of A_{day}, and the error was larger with higher LAI (i.e., the M-S$_2$ model vs the Y-plant model). Consequently, A_{day} could be overestimated by an average of 213% as a result of ignoring dynamic responses of light. It is necessary to understand the possibilities and limitations of the proposed model and to determine which models make more efficient use of survey data. The application of such models often depends on the availability and quality of required data. The quantitative, comparative studies are important to gain a better understanding of physiological process. Our study showed the possible ranges how much A_{day} was affected by light distribution patterns, stand density and seasonal changes in the light environment.

Chapter 4 - Remote sensing has facilitated extraordinary advances in modeling, mapping, and the understanding of ecosystems. Applications of remote sensing involve either images from passive optical systems, such as Aerial Photography and the Landsat Thematic Mapper, or, active Radar sensors such as RADARSAT. These types of remote sensors have proven to be satisfactory for many forest applications, such as mapping and classifying land cover into specific classes and, in some biomes, estimating aboveground biomass and Leaf Area Index (LAI). However, conventional sensors have significant limitations for ecological and forest applications. The sensitivity and accuracy of these devices have repeatedly been shown to fall with increasing aboveground biomass and LAI. They are also limited in their ability to represent the spatial patterns. They produce only two-dimensional (x and y) images, which cannot fully represent the three dimensional structure of the forest canopy. Ecologists have long understood that the presence of specific organisms and the overall richness of wildlife communities can be highly dependent on the three-dimensional spatial pattern of vegetation. Individual bird species, in particular, are often associated with specific three dimensional features in riparian forests. Additionally, aspects of forests, such as productivity, may be related to forest canopy structure.

Lidar (light detecting and ranging) is an alternative remote sensing technology that promises to both increase the accuracy of biophysical measurements and extend spatial analysis into the third dimension (z). Lidar sensors directly measure the three-dimensional distribution of forest canopies as well as sub-canopy topography, therefore providing high resolution topographic maps and highly accurate estimates of tree height, cover, and canopy structure. In addition, lidar has been shown to accurately estimate LAI and aboveground biomass, even in those high biomass ecosystems, where passive optical and active radar sensors typically fail to do so. Estimation of forest structural attributes, such as LAI, is an important step in identifying the amount of water use in forest areas.

Chapter 5 - Soil respiration, CO_2 efflux from the soil surface, is an important process of the carbon (C) cycle in terrestrial ecosystems. Soil respiration includes many processes involving biotic factors, such as respiration from roots and microorganisms, along with abiotic factors and various temporal and spatial factors. Many researchers have examined soil respiration of various ecosystems. Recently, tropical forests have been converted into secondary forests or agricultural forests. This land use change might strongly affect the global C cycle; nevertheless, few data are available to reflect land use effects on dynamics of soil organic carbon (SOC) in Southeast Asia. Therefore, we established study sites at four different ecosystems (primary forest, secondary forest, oil palm plantation, and rubber plantation) in the Pasoh area of Malaysia in Southeast Asia. This study was designed to determine spatial and temporal variations (diurnal and seasonal change) of the soil respiration rate and to estimate the annual C efflux from soil and dynamics of SOC in different ecosystems.

Seasonal data suggest that the soil respiration rate is negatively correlated with soil water contents in the primary forest, secondary forest, and rubber plantation. The soil water content shows a negative correlation with the gaseous phase content. The gaseous phase content shows a positive correlation with soil respiration rate at all sites.

The annual C efflux from soil was estimated as 16.9–19.2 t C ha^{-1} in the primary forest, 17.5–18.5 t C ha^{-1} in the secondary forest, 14.3–14.5 t C ha^{-1} in the oil palm plantation, and 9.0–11.2 t C ha^{-1} in the rubber plantation. Moreover, we estimated the annual SOC budgets using the three-box model. Results suggest that the biomass of dead roots, turnover time, and contribution of heterotrophic respiration are important factors for accurate evaluation of soil C dynamics and budgets.

Chapter 6 - Plants growing under dense canopy experience conditions - low light intensity, recycling of biogenic CO_2 - that lead to lower $^{13}C/^{12}C$ ratios than plants growing in open landscapes. This particular stable isotopic ratio is passed on tissues of herbivorous mammals feeding under dense canopy conditions, including their bone collagen. As bone collagen can be extracted from ancient bones several thousand years old, its carbon isotopic signature can be used to track the development of dense forest through time. One good example is the development of dense forest in France during the last 35,000 years, especially since the beginning of the Holocene about 10,000 years ago. Using dated large bovid bones and teeth (from Bison *Bison priscus* and Aurochs *Bos primigenius*) from archaeological sites as tracers of dense forest development, it appears that open conditions dominated from 35,000 to 10,000 years ago, although changes in the vegetation composition were observed. In contrast, a very significant decrease of $^{13}C/^{12}C$ ratio during the Preboreal period, around 10,000 years ago, seems to correspond to the spread of dense temperate forest in France and is directly linked to climatic change. A further decrease of $^{13}C/^{12}C$ ratio in aurochs bone collagen occurs during the Late Atlantic (6,000-5,000 years ago), which coincides with the extension of agricultural societies and the development of cattle husbandry. This further decrease seems to correspond to some clearance of the dense forest cover for agricultural purpose and particularly to provide feeding grounds for cattle, while the wild aurochs tend to live in deep dense forest and avoid forest edges and open environments. When local situations are explored further, for instance the Paris area and the French Jura during the Middle Neolithic, regional differences are found regarding the use of dense forest biomass by prehistoric farmers. This new approach is a promising one that could allow fine level tracking of dense forest exploitation by humans through time.

Chapter 7 - The ongoing anthropogenic climate change is immensely altering structure and function of the terrestrial biosphere, including forest ecosystems. In turn, the changing ecosystems have a strong potential to modify the climate through changes in biogeochemical cycles (e.g., C storage) and biophysics (e.g., albedo and hydrological cycling). Forest ecosystems will have particularly significant impacts onto the climate due to their large terrestrial coverage, vast C stock, and prominent biophysical characteristics. To reproduce the two-way interaction between vegetation and climate, climate models should be integrated with dynamically responding vegetation models. Here we present our recent progress, concerns, and future directions in simulations of vegetation processes by the terrestrial biosphere model (TBM) sSEIB (a simplified version of SEIB-DGVM) that is coupled to a climate system model (Center for Climate System Research-Frontier Research Center for Global Change general circulation model, CCSR-FRCGC GCM). sSEIB explicitly reproduces the ecophysiological, population, and community dynamics based on an individual-based forest model representation. The model is also fully coupled to the global biogeochemical cycling that in turn affects atmospheric CO2 concentrations. The GCM-coupled sSEIB successfully reproduced the current global distributions of vegetation types and plant production. A preliminary climate change experiment with the stand-alone sSEIB showed significant responses of terrestrial vegetation and soil C storage.

Chapter 8 - Forest canopy may be an important interface between atmosphere and forest ecosystems. Properties of leaf surface as a major part of the forest canopy were studied mainly in Japanese cedar (*Cryptomeria japonica*) forests. The amount of epicuticular wax increased under the effect of water stress (on high branches and at locations with low rain factors), exposure to the noxious gases (such as volcanic acidic gases), and strong UV radiation at high altitude, while the C content of wax decreased and the O content increased, except in case of the altitude. In the Kanto Plain around Tokyo, where Japanese cedar is declining, the wax eroded more rapidly (approximately 1.5 times faster) than that of healthy trees in mountainous areas, although the amount of wax in current-year leaves was almost equivalent in both areas. Amounts of anthropogenic elements such as antimony (Sb) in particulate matters deposited on the leaf surface were greater (10 times greater in case of Sb) at the severe decline area than at the healthy area. Atmospheric deposition including the particulate matters may be a possible cause of the wax degradation. Fractions of unhealthy stomata (disability in closing) correlated with the amounts of particulate Sb on the leaves. In fact, stomata clogged with particles were observed by optical or scanning electron microscopes. The cuticular transpiration rate in 1-year leaves was higher in the plain area (0.92% h^{-1} as decreasing rate of leaf water content) than in the mountainous area (0.60% h^{-1}). This water loss, resulting from a degraded wax layer and partial malfunctioning of stomata due to deposited particulate matters, may be a significant factor causing the decline of Japanese cedar. Recent dry atmospheric conditions in the plain area may accelerate the water stress of trees. Anthropogenic elements in the particulate matters on the leaf can be utilized as indicators of air pollution, too. The amounts of Sb correlated with NO_X concentration and population density in each sampling area. Moreover, epicuticular wax properties and particulate matters on the leaf may affect leaf surface wettability. Increase in leaf wettability may accelerate ion exchange on the leaf surface, resulting in increase of leaching of K^+ and uptake/consumption of N compounds on the forest canopy. The canopy interactions should be considered for discussion of elemental flows in forest ecosystems. Small changes in leaf surface properties on the forest canopy may affect plant physiology and biogeochemical cycles of elements in forest ecosystems.

Chapter 9 - This study explored the effects of canopy gaps and gap size on leaf litter microarthropod abundance at five sites within the experimental forest of the Coweeta Hydrologic Laboratory, in the Nantahala Mountain Range of western North Carolina. In March, 2002, five canopy gaps were formed by pulling over trees to mimic natural disturbances. Two of the gaps measured 20m in diameter and three measured 40m. At each of the five gap sites we placed four experimental plots under canopy gaps and two control plots under closed canopy located just outside the gaps. Microarthropods were collected using $15cm^2$ mesh bags filled with two grams of leaf litter. Litter bags were placed in a three-by-four pattern at each plot on February 1, 2004. Litter bags and soil data were collected every other month for two years. A generalized linear mixed model was used to investigate effects of canopy gap presence and size on the populations of collembola, oribatid mites, prostigmatid mites, and mesostigmatid mites. Microarthropod counts were typically significantly greater in control plots than in gaps, with no significant effect due to gap size.

Chapter 10 - Within the framework of micrometeorology and biosphere-atmosphere interactions, one of the important scientific tasks has recently been to conduct long-term flux measurement sites in an array of land biomes and climates world-wide in order to gain understanding of exchange processes with the good spatio-temporal coverage. At the moment, the flux network is rather dense for natural and seminatural ecosystems while only a few measurements have been conducted over urban and sub-urban landscapes. Urban vegetated surfaces do not likely contribute much to overall material balances, like carbon sinks/sources, but locally they may be important; they have a significant role in the well-being of the population, and knowledge of them is scarce. Besides carbon, vegetation also affects the energy balance by modifying surface temperature. Canopies may also act as significant sinks for aerosol particles thus cleaning the air, although vegetation can also produce particles. Apart from telling the impacts that urban vegetation have on the local air properties, the flux studies provide valuable information on the performance of the vegetation, which is valuable knowledge for the maintenance of green surfaces in urban areas. We discuss the present status of urban flux studies from the point of view of vegetated surfaces and try to point out areas that need to be explored more. A few illustrative results on flux studies from Helsinki, Finland, are presented

In: Forest Canopies: Forest Production, Ecosystem...   ISBN 978-1-60741-457-5
Editor: J. D. Creighton and P. J. Roney   © 2009 Nova Science Publishers, Inc.

Chapter 1

A New Paradigm of Forest Canopy Interception Science: The Implication of a Huge Amount of Evaporation During Rainfall

Shigeki Murakami

Tohkamachi Experimental Station, Forestry
and Forest Products Research Institute, Tohkamachi, Japan

ABSTRACT

Canopy interception includes some enigmatic problems. Firstly, evaporation rate is proportional to the rainfall intensity. Secondly, the heat budget model underestimates the measured canopy interception at some sites but the Gash models do not, though both types of models are physically based. Thirdly, though high evaporation rate during rainfall, e.g. ≥ 10 mm h^{-1}, was observed, the source of latent heat of vaporization and efficient water vapor transfer mechanism during rainfall remain unsolved. The first problem, the dependence of canopy interception on rainfall intensity, DOCIORI, cannot be explained by the conventional heat budget model, since rainfall intensity is not included as a parameter. To elucidate the phenomenon, the idea of splash droplet evaporation, SDE, was proposed. As the raindrop size and the number of raindrops per unit volume increase with the rainfall intensity, so does the number of splash droplets produced by the raindrops hitting on the canopy. Splash droplets evaporate efficiently due to large combined surface area. The DOCIORI was pointed out in the Japanese sites for the first time in the 1980s, but has not been known worldwide until recently. In this chapter it is shown that the DOCIORI occurs at sites other than in Japan using literature data in which authors did not notice the phenomenon. The second problem is also understandable by SDE; the heat budget model considers evaporation from canopy surface only but actual evaporation occurs from the surface of splash droplets as well. This overlook leads to underestimation of the heat budget model especially where rainfall intensity is sufficiently high, while estimation by the Gash models reproduces the measurement fairly well, because they include SDE implicitly. The third problem concerns not only forest hydrology but also atmospheric sciences, and is explainable at least qualitatively by the innovative idea of evaporative force proposed recently, which is an unnoticed basic principle in meteorology.

INTRODUCTION

Evaporation of canopy interception is seemingly a simple phenomenon. Rain water on the canopy surface evaporates and, consequently, less rain water reaches the forest floor than on the ground at an open site. Canopy interception ranges from 9% to 48% of annual rainfall (Hörmann et al., 1996), and is a major component of evaporation from the forest. The rest of the water reaches the forest floor and can be utilized by living things in the watershed. Regardless of the importance of canopy interception in terms of water budget, the mechanism remains unsolved; it includes two major enigmatic problems.

The first problem is that evaporation of canopy interception is proportional to the rainfall intensity. Though this phenomenon was pointed out in the 1980s for the first time, unfortunately, it has been known only in Japan as the papers were written in Japanese (Hattori et al., 1982; Tsukamoto et al., 1988). The second problem concerns the heat budget. In some areas most evaporation occurs during rainfall not after the cessation of the rain. It sounds unusual that evaporation occurs during rainfall, but a more important thing relevant to this is that the source of latent heat of vaporization and the sink of water vapor are a mystery as well as the transport mechanism of them. In some cases the hourly evaporation rate of the forest canopy interception exceeds 10 mm h^{-1} (Hashino and Tamura, 2005) which requires latent heat of 6814 W m^{-2} at 20°C. This value corresponds to five times the solar constant. Relative humidity during a rain event seldom reaches 100%, only around 95%, and that means water vapor is removed by a certain process even on the condition that the entire ground surface gets drenched and there are numerous rain drops in the air. From where is latent heat supplied? Why does water vapor not saturate during rainfall? In a sense, those questions are beyond the range of forest hydrology and more in the area of atmospheric sciences. I believe forest hydrologists have noticed the contradiction. However, they have been able to do nothing to solve the problem. Many papers on canopy interception state that latent heat is supplied by "advection" (Stewart, 1977; Singh and Szeicz, 1979; Pearce et al., 1980; Bruijnzeel et al. 1987; Dykes, 1997; Schellekens et al. 1999; Waterloo et al. 1999; van der Tol et al., 2003; Wallace and McJannet, 2006; 2008). Advection means horizontal transport of a warm air mass to the observation site, and it must be detectable if micrometeorological measurements are conducted; unfortunately, no such data have ever been shown where the measurements were made. Those two cardinal problems have remained unsettled for decades since there seemed to be no clue to the solution

The first problem is explainable by the idea of splash droplet evaporation, SDE, proposed recently (Murakami, 2006), though the details of the phenomenon remain unsolved. In regard to the second problem, a heat budget approach that requires micrometeorological data, e.g. Rutter type model (Rutter et al., 1971; 1975), does not seem to be applied so often lately. Instead, the Gash models (Gash, 1979; Gash et al. 1995) prevail; a major reason is probably that at some sites the heat budget method does not fit measured data very well, even if the parameters are optimized (the section *Reconsideration on the heat budget model* in DISCUSSION). On the other hand, the Gash models can well reproduce measured data overall, though the difference between the heat budget approach and the Gash models has not been understood.

There are three purposes for this chapter, all of which will contribute to solving the above-mentioned problems. The first is to review the idea of SDE based on the measurement conducted in a young stand of Japanese cypress and on literature data, other than that from Japan, that include SDE implicitly (the section THE IDEA OF SDE), which is a basis to explore the following two points. The second is to clarify the relation among the canopy interception models focusing on SDE. Three canopy interception models—the heat budget, the revised Gash and the newly constructed SDE model—are applied to the young stand of Japanese cypress to elucidate their differences and their common features (the section INTERRELATION OF THE MODELS). This helps provide a deeper understanding of the phenomenon. The third is to review an innovative idea that would solve the heat budget contradiction (A CLUE TO SOLVE ENORMOUS EVAPORATION; based on Makarieva and Gorshkov, 2007). It is an overlooked basic principle in meteorology, and is applicable to not only canopy interception but meteorology in general.

THE IDEA OF SDE

The idea of SDE is exemplified in this section based on observed canopy interception in a Japanese site and literature data obtained from sources other than Japan.

SDE in a Young Stand of Japanese Cypress

Canopy interception was measured in a young stand of Japanese cypress in the Hitachi Ohta Experimental Watershed, Japan. The detail of the site is described in the next section SITE AND INSTRUMENTATION.

The relation between hourly rainfall (rainfall intensity) and hourly canopy interception is analyzed for a rain event of 20 mm or more (Figure 1). The separation time of the rain event is set at 6 hours (SITE AND INSTRUMENTATION). The threshold rainfall amount of 20 mm was determined, because at least three data, i.e. three-hour duration, are required to calculate a regression line and most rain events larger than or equal to 20 mm satisfies this condition. Hourly canopy interception is clearly proportional to the rainfall intensity, and the inclination of the regression line is defined as DOCIORI (dependence of canopy interception on rainfall intensity) or i. As described in the section "THREE CANOPY INTERCEPTION MODELS", this phenomenon cannot be explained by evaporation from canopy surface only, and the idea of SDE was proposed by Murakami (2006) to explain the DOCIORI.

As the raindrop size and the number of raindrops per unit volume increase with rainfall intensity (Marshal and Palmer, 1948), so does the number of splash droplets produced by raindrops hitting the canopy. Efficient evaporation occurs from the numerous small droplets that have a huge combined surface area. As a result, evaporation of canopy interception increases with rainfall intensity.

The simulation under relative humidity of 95% demonstrated that a droplet with a radius of 25 μm evaporates and completely disappears with a fall of 1.7 m at 10 °C and 2.8 m at 25 °C. While a droplet of 100 μm in radius reduces only 2.4% and 3.7% of the initial mass at 10°C and 25 °C, respectively, with a fall of 8 m. The value of relative humidity, 95%, represents actual relative humidity during rainfall; hourly mean relative humidity from 1999 to 2000 is roughly constant through the year at the Mito Meteorological Observatory (25 km southwest of Hitachi Ohta), namely, 94% with a standard deviation of 1.9. The simulated results are very sensitive to the droplet size, and you can understand this result as follows. This is the same principle with a spray; if you pour a glass of water on the floor, it is drenched and puddles are formed. While if you spray the same amount of water in the air using an atomizer, the floor gets wet but puddles are not formed because sprayed water evaporate rapidly during the fall. Sprayed or splash droplets have large surface area relative to mass that strongly promotes evaporation with decreasing the droplet size.

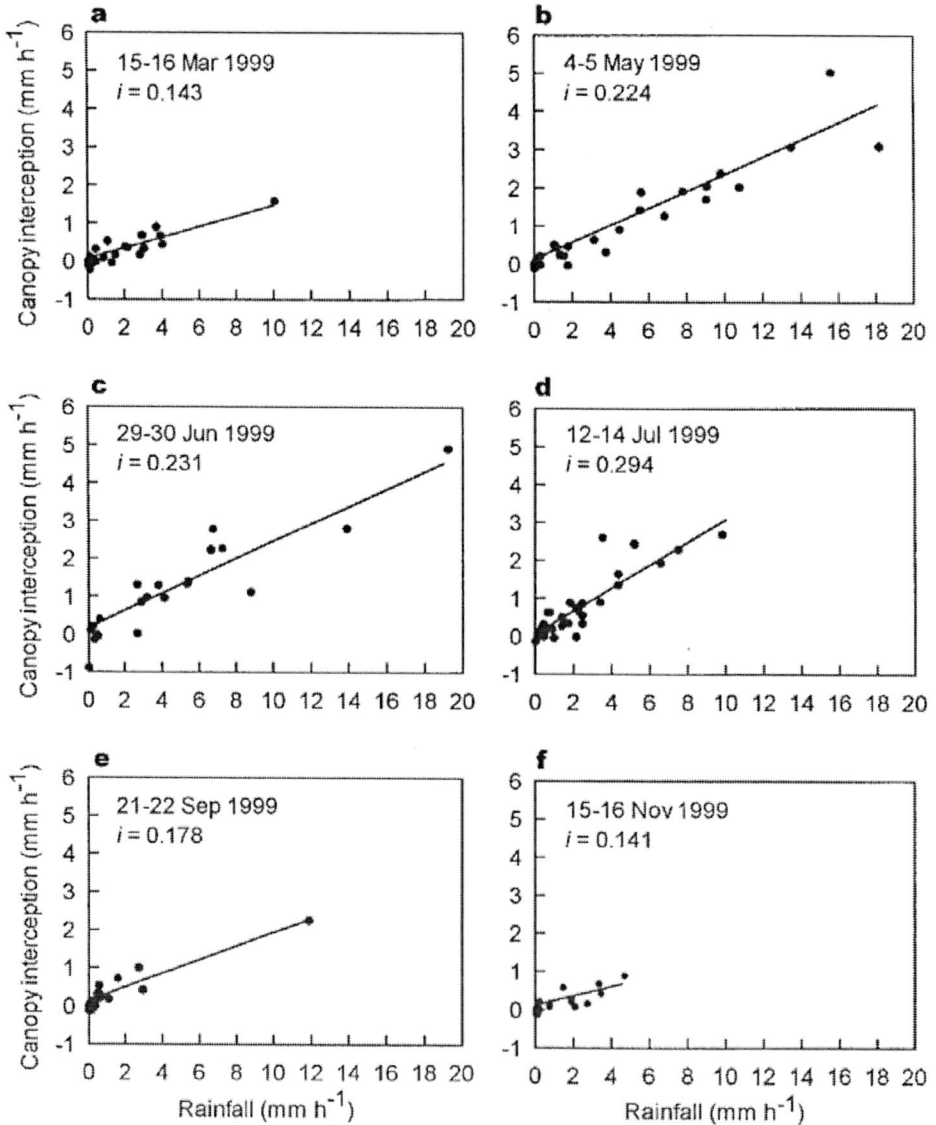

(Reproduced from Murakami, 2006; Copyright Elsevier B.V.)

Figure 1. Hourly rainfall and hourly canopy interception. The inclination of the regression line is defined as DOCIORI, i. A value of i is shown in each panel. Note that the values of i vary with seasons, and that some values of canopy interception are negative due to measurement error. The regression lines are calculated excluding these negative values. Total rainfall of each panel: (a) 40.2 mm, (b) 132.6 mm, (c) 94.3 mm, (d) 72.3 mm, (e) 22.5 mm, (f) 24.2 mm.

Figure 1 evidently shows seasonal variations in i with larger values in summer. All the data of i for 1999 to 2000 is superposed and shown in Figure 5 that has a peak in summer. This can be partly explainable as follows. As mentioned above relative humidity is constant through the year, and this means water vapor deficit is a function of air temperature only. High water vapor deficit in summer facilitates evaporation from both the canopy surface and the surface of the splash droplets that coincides with variations in i. However, the amount of seasonal changes in evaporation by this process is not enough to explain measured values quantitatively (Murakami, 2006). Other processes, e.g. seasonal changes in LAI and seasonal variations in the raindrop size distribution, might govern seasonal changes in i. For further study we need to know the splash droplet size distribution and dependency of splash droplet productivity on raindrop size and tree species that might also change with season.

SDE Other than Japanese Sites

As described below, Murakami (2007a) pointed out that the DOCIORI had been detected in several studies, though the authors did not notice it. This implies that the DOCIORI and SDE occur all over the planet where the rainfall intensity is sufficiently high.

Rainfall and throughfall data obtained in a tropical rainforest in Puerto Rico are presented at a 5-minute interval in Schellekens et al. (1999). They did not measure stemflow because it was negligible small at the site. Their data enable to analyze the relation between hourly rainfall p_g and hourly canopy interception I_h, that is, DOCIORI. Figure 2 (a) – (d) were drawn based on Figure 3 (a) and (c), and Figure 4 (a) and (c) in Schellekens et al. (1999), respectively. DOCIORI is apparent in Figure 2 (a) – (c), though correlation coefficient shows low value in Figure 2 (d). The values of i in Figure 2 (a) – (c) (0.478 to 0.625) are much higher than those of Figure 1 (0.092 to 0.294). The cause of this might be the difference in rain drop size distribution between the sites for the same rainfall intensity (dependence of Marshall-Palmer distribution on sites) in addition to meteorological factors and forest architectures.

Llorens et al. (1997) observed canopy interception of *Pinus sylvestris* in a Mediterranean mountain region in Spain that also reveals DOCIORI, though the data were not shown on an hourly basis but on a rain event basis. Their Table 1 lists vapor pressure deficit (VPD), rainfall duration, rainfall amount, and canopy interception. Mean hourly rainfall $\overline{P}_G$ (rainfall amount divided by rainfall duration) and mean canopy interception for the rain event $\overline{I}$ (canopy interception divided by rainfall duration) were derived from the data. The relation between $\overline{P}_G$ and $\overline{I}$ is displayed in Figure 3 (a) for all the data, VPD < 1 hPa, and VPD ≥ 1 hPa with regression lines. $\overline{I}$ increases with $\overline{P}_G$ with fairly large correlation coefficients that means DOCIORI, though scatter in data is larger for VPD ≥ 1 hPa. The inclination of the regression lines is not i per se but indicator of i since $\overline{I}$ and $\overline{P}_G$ are not hourly data but the mean values over the rain event. The maximum $\overline{P}_G$ for this dataset was 13.0 mm h^{-1}.

Canopy interception measured in a seasonal temperate rainforest of old-growth Douglas-fir (> 450-year-old) in southwestern Washington, UAS, by Link et al. (2004) also showed the DOCIORI, though the correlation coefficient is not high, 0.584 (Figure 3 (b)). The maximum $\overline{P}_G$ was only 2.1 mm h^{-1}, and this low rainfall intensity might be one of the reasons for small correlation coefficient. The subsequent work conducted by Pyker et al. (2005) at the 25-year-old Douglas-fir stand that is some 4 km away from the old-grown Douglas-fir forest (Link et al. 2004) also posts I and P_G on a rain event basis with the duration of the rainfall. The regression line is expressed as, $\overline{I} = 0.172\overline{P_G} + 0.098$, $r=0.728$.

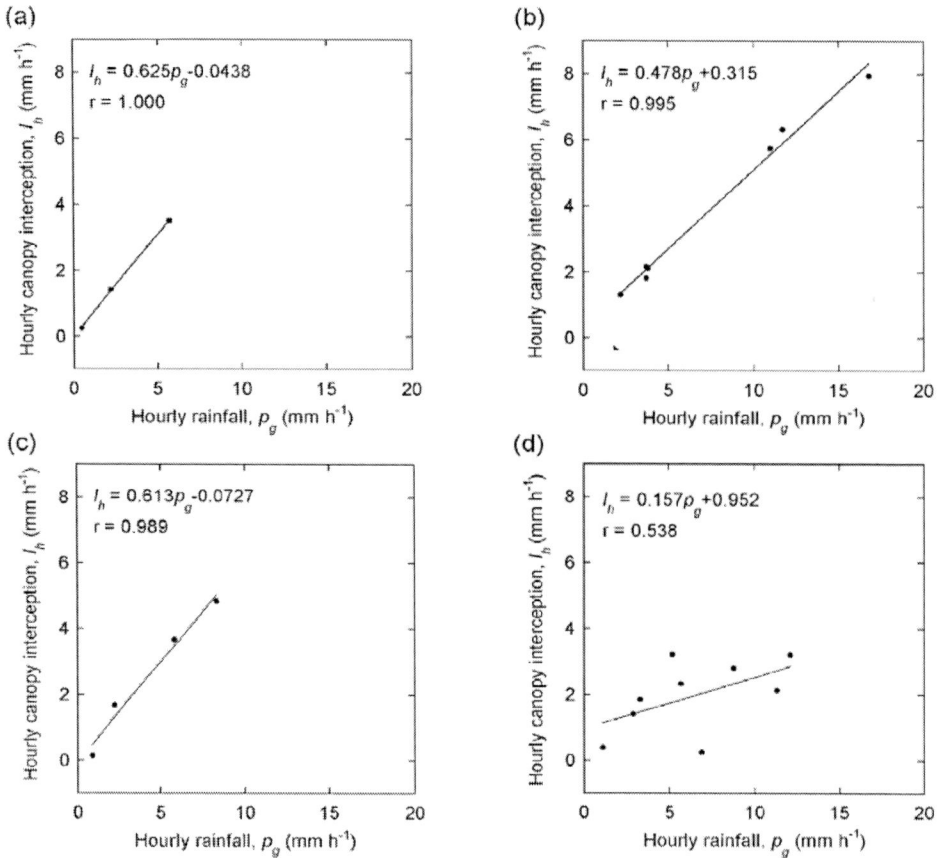

(Reproduced from Murakami, 2007a; Copyright Elsevier B.V.).

Figure 2. Hourly rainfall p_g and hourly canopy interception I_h, that is, DOCIORI, i, in a tropical rainforest in Puerto Rico. The panel (a) was dawn based on Figure 3a in Schellekens et al. (1999), (b) Figure 3c in Schellekens et al. (1999), (c) Figure 4a in Schellekens et al. (1999), (d) Figure 4c in Schellekens et al. (1999). The variables in the hozirontal and the vertical axis, p_g and I_h, are equivalent to those in Figure 1. Note that the inclinations of the regression lines except for (d) are much steeper than those of Figure 1.

The number of data is only eight and much fewer than those of Link et al. (2004) shown in Figure 3 (b), while the maximum $\overline{P_G}$ was 3.5 mm h^{-1} that is higher than the value of Link et al. (2004), 2.1 mm h^{-1}.

The correlation coefficients are higher and the scatter in data are smaller in Figure 1 (0.844 – 0.925; Murakami, 2006) and Figure 2 (0.538 – 1.000; Schellekens et al. 1999) than those in Figure 3 (a) (0.724 – 0.778; Llorens et al., 1997), (b) (0.584; Link et al. 2004), and Pyke et al. (2005), 0.728. Figure 1 (Murakami, 2006) and Figure 2 (Schellekens et al. 1999) that use hourly data, p_g and I_h, represent the DOCIORI, while Figure 3 (a) (Llorens et al., 1997), (b) (Link et al. 2004), and Pyke et al. (2005) are compiled using the mean values over a rain event, $\bar{I}$ and $\bar{P}_G$, which degrade the correlation coefficients since the rainfall intensity is reduced for longer period of average. Even so, these data show that the DOCIORI occurs in Spain and the USA in addition to Japan and Puerto Rico.

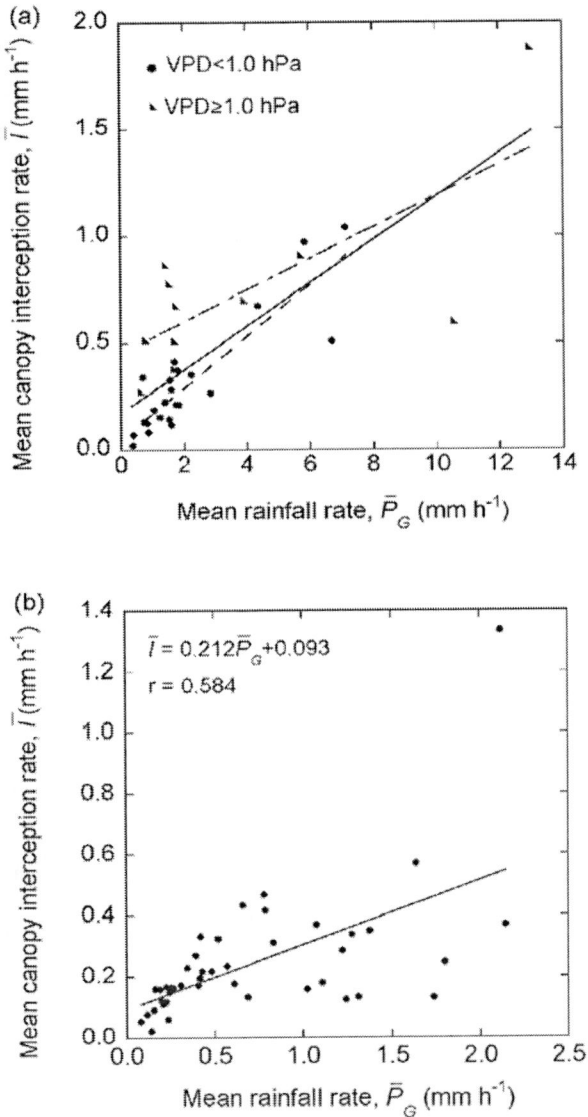

Figure 3. DOCIORI for sites in Spain and the USA. Each individual data (circle or triangle) corresponds to a single rain event. (a) *Pinus sylverstris* in a Mediterranean mountainous area of Spain. Data are based on Table 1 of Llorens et al. (1997). Three regression lines are shown. Soil line: for all data; $\bar{I} = 0.101\bar{P}_G + 0.179$, $r=0.778$. Dash–dot line: for VPD$\geq$ 1.0 hPa; $\bar{I} = 0.0737\bar{P}_G + 0.452$, $r=0.724$. Dashed line: for VPD<1.0 hPa; $\bar{I} = 0.121\bar{P}_G + 0.051$, $r=0.881$. (b) Douglas fir in the USA. Data are derived from Table 3 in Link et al. (2004).

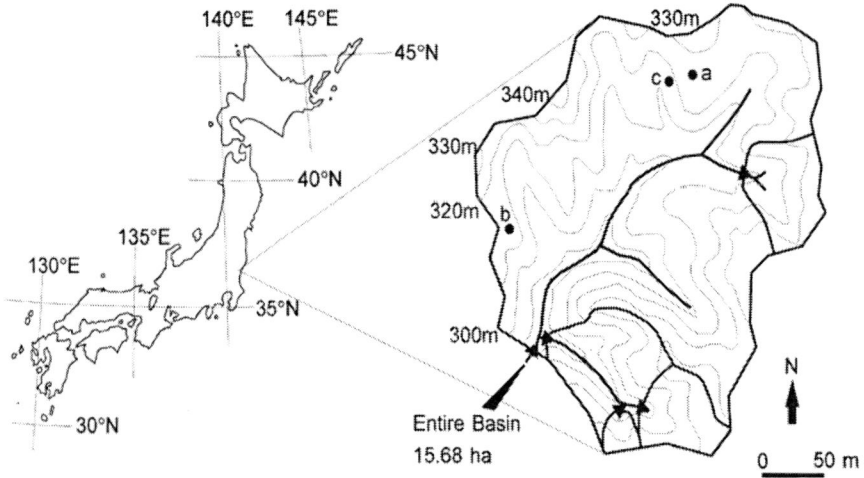

(Reproduced from Murakami, 2006; Copyright Elsevier B.V.)

Figure 4. Hitachi Ohta Experimental Watershed. (a) Meteorological station with a set of raingauge, (b) a set of raingauge, (c) canopy interception plot, and gauging weir (▲).

Table 1. Monthly precipitation in the Hitachi Ohta Experimental Watershed. Rainfall amount used for the analysis is less than the values shown in this Table, because snowfall, occult precipitation, and negative canopy interception caused by error were excluded (see RESULT)

Year	1999	2000	Average 1991-2000
	mm	mm	mm
January	2.0	71.6	51.6
February	41.5	19.3	49.6
March	135.5	74.8	112.5
April	304.5	186.5	127.6
May	203.0	181.8	186.3
June	250.9	187.7	165.6
July	276.0	249.1	164.7
August	135.2	45.7	127.5
September	100.8	226.2	219.7
October	173.6	125.5	135.4
November	67.6	77.9	94.5
December	17.3	6.8	32.7
Total	1707.9	1452.9	1467.7

(Revised and reproduced from Murakami, 2006; Copyright Elsevier B.V.)

SITE AND INSTRUMENTATION

The study on SDE in the previous section, *SDE in a young stand of Japanese cypress* in THE IDEA OF SDE, was conducted in the Hitachi Ohta Experimental Watershed, the Pacific Coast of eastern Japan, 36°34' N, 140°35' E (Figure 4). The data were also used for the three canopy interception models in the next section. The Watershed was established in 1906 and once closed in 1919, but the measurement resumed in 1980. During the history natural forest was clear-felled followed by plantations to evaluate the effect of forest changes on runoff. After the resumption not only water budget but also process studies on forest hydrology have been conducted in the Watershed (Sidle et al., 1995; 2000; Tsuboyama et al. 1994; 2000). The research described in this chapter was conducted in a young stand of Japanese cypress, *Chamaecyparis obtusa*, for two years 1999 to 2000 (Murakami, 2006; 2007a; 2007b). Monthly precipitation and characteristics of the stand are shown in Table 1, and Table 2, respectively. Table 1 shows that winter is dry season with small precipitation. Leaf area index (LAI) measured by the LAI-2000 Canopy Analyzer (LI-COR Inc. Lincoln, NE, USA) revealed seasonal change with a peak in summer (Table 2).

Five gauging weirs, two sets of raingauges, a weather station, and a canopy interception plot were located as shown in Figure 4. Using the instruments manufactured by Ikeda Keiki Seisakusyo, rainfall, global radiation (SR-180), wind speed and wind direction (KS82P), air temperature, and relative humidity (HR 202S, thermistor and film polymer sensor) were measured at the meteorological station. The raingauge at the meteorological station, the point *a* in Figure 4, was 0.1 mm per tip recorded at a 10-minutes interval, while at the point *b* one tip was 0.5 mm recorded hourly. Both the points are located at the ridge, and vegetation around the gauges was cut so that it does not affect rainfall measurements. Those tipping bucket raingauges were set on the ground level, and the storage type raingauges that were measured every two weeks were also installed at both sites. The rainfalls measured by the storage type gauges were regarded as correct values, and those of the tipping bucket gauges were modified.

The separation time of the rain event was 6 hours, i.e. the rainfall is regarded as a discrete rain event when rainfall was not observed for 6 hours or more after the cessation of the rainfall.

Table 2. Characteristics of the young stand of Japanese cypress, Chamaecyparis obtusa

Year		1999	2000
Age		11	12
Average tree height	m	5.8	6.3
Average diameter at breast height (DBH)	cm	7.0	8.1
Stand density	stem/ha	2944	2944
Leaf area index (LAI)	ha/ha	3.7 – 4.3*	3.9 – 5.2*

*Seasonal variation is evident.

For a rain event of more than or equal to 5.0 mm, mean values of the two tipping raingauges was adopted, while for that of less than 5.0 mm the value measured by the raingauge with 0.1 mm per tip at the point a was used.

At the canopy interception site throughfall and stemflow were measured (the point c in Figure 4). Canopy interception was calculated as the difference between gross rainfall measured at the point a and/or b and net rainfall (= throughfall + stemflow). Throughfall was collected using two troughs 590 cm long and 18 cm wide, while stemflow collectors were set on nine trees (Figure 3and 4 in Murakami, 2008). In the previous publications the number of stemflow collectors was mistakenly described as eight. The flaky bark of $C.$ $obtusa$ was shaved to smooth, and the stemflow collector made of a polyurethane board that somewhat absorbs the growth of the trunk was fastened around the stem using plastic cords. A gap between the board and the bark was filled with silicone sealant. Coarse plastic sponge was placed at the outlet of the trough and the stemflow collector to avoid getting plugged by litter. Throughfall and stemflow flow into each individual tank through plastic tubes, and the water level of each tank was recorded on the chart paper with the resolution of 0.1 mm and 0.07 mm, respectively. The tank siphons out water automatically if the water level reaches maximum (Figure 5 in Murakami, 2008).

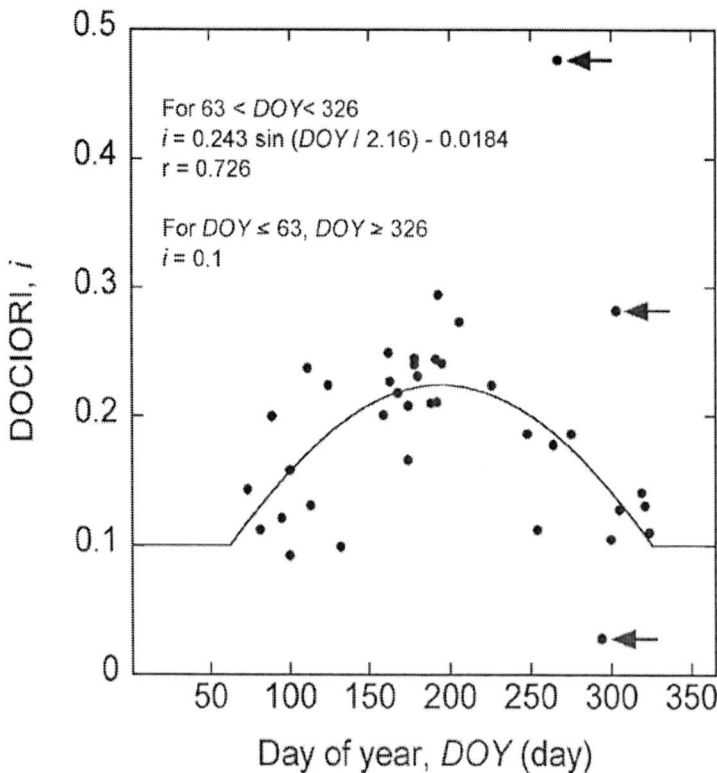

For $63 < DOY < 326$
$i = 0.243 \sin (DOY / 2.16) - 0.0184$
$r = 0.726$

For $DOY \le 63$, $DOY \ge 326$
$i = 0.1$

(Reproduced from Murakami, 2007a; Copyright Elsevier B.V.)

Figure 5. DOY (day of year) and DOCIORI (dependence of canopy interception on rainfall intensity). DOCIORI, i, is defined as an inclination of a regression line between hourly rainfall and hourly canopy interception, e.g. the regression line in Figure 1. Two years of data from 1999 to 2000 are plotted in the Figure. The regression curve is calculated except arrowed data that are outlier. The unit of the sine function is degrees.

THREE CANOPY INTERCEPTION MODELS

Three canopy interception models, the conventional heat budget model (Murakami et al., 2000), the SDE model that is referred to as the "DOCIORI model" in Murakami (2007a), and the revised Gash model (Gash et al., 1995) are applied to the young stand of Japanese cypress in the Hitachi Ohta Experimental Watershed. The heat budget model and the SDE model run under the common backbone model with the different evaporation terms as described in each section below.

Backbone Model

The backbone model based on Kondo et al. (1992) calculates evaporation of canopy interception on a rain event basis considering one or two evaporation processes depending on a storm size. For a small rain event that satisfies the inequality $c^*P_G < E\tau + S_{max}$ shown in Eq. (1), evaporation rate from the canopy and the water storage on the canopy and the trunk are larger than the capture rate of raindrops. All the rainwater caught by the canopy evaporates, and canopy interception I is written as,

$$I = c^* P_G \qquad \text{for } c^* P_G < E\tau + S_{max} \tag{1}$$

where c^* is a chance of a raindrop hitting the canopy, P_G is gross rainfall, E is evaporation rate that is estimated using the heat budget model or the SDE model described later, τ is the duration of rain event. S_{max} is the sum of water capacity on the canopy S and on the trunk S_t.

For a large storm that satisfies the inequality $c^* P_G \geq E\tau + S_{max}$ demonstrated in Eq. (2), the two interception processes should be considered; evaporation during rainfall and after the cessation of the rainfall.

$$I = E\tau + S_a \qquad \text{for } c^* P_G \geq E\tau + S_{max} \tag{2}$$

where S_a is the total water stored in the canopy and the trunk after the end of the rainfall. The value of S_a increases with P_G due to wetting-up the canopy, and for longer duration of rainfall the trunk also gets wet. Consequently, S_a approaches S_{max}. In the same way, c^* approaches the figure of canopy closure c ($0 \leq c \leq 1$) with increasing P_G. It is assumed that the value of c^* also increases with leaf area index, LAI, and leaf inclination factor, F (assumed to be 0.5). Based on these consideration S_a and c^* are presumed to be written as,

$$S_a = S_{max}\left[1 - exp\left(-\frac{P_G}{S_{max}}\right)\right] \tag{3}$$

$$c^* = c\left[1 - exp\left(-F\frac{LAI}{c}\right)\right] \tag{4}$$

S was estimated from the relation between P_G and throughfall assuming minimal evaporation during rainfall (Leyton et al., 1967), and S_t was obtained by the relation between P_G and stemflow. LAI is assumed to be constant (4.5) based on the measurement from 1999 to 2000, and c was measured by a tree survey. The figures of S, S_t and c were 0.41 mm, 0.19 mm, and 0.81 in 1999, and 0.44 mm, 0.13 mm, and 0.94 in 2000, respectively.

Heat Budget Model

Evaporation rate E in Eqs. (1) and (2) are calculated based on the conventional heat budget equation and the transfer equations.

$$R_n = H + \lambda E + G \tag{5}$$

R_n is net radiation, H is sensible heat, λ is latent heat of vaporization, and G is ground heat flux. The downward long wave radiation was estimated using air temperature, relative humidity, and sunshine duration (Kondo et al. 1991). Albedo and G were assumed to be 0.1 and zero, respectively. The transfer equations for sensible heat and latent heat are,

$$H = \frac{c_p \rho}{r_a} (T_s - T_a) \tag{6}$$

$$E = \frac{\rho}{r_a} (q_{sat}(T_s) - q(T_a)) \tag{7}$$

where c_p is the specific heat of air at constant pressure, ρ is air density, T_s is surface temperature, T_a is air temperature, q_{sat} is saturated specific humidity, and q is specific humidity. r_a is aerodynamic resistance described as follows,

$$r_a = \frac{1}{\kappa^2 u} \left[ln\left(\frac{z-d}{z_0} \right) \right]^2 \tag{8}$$

where κ (= 0.4) is von Karman's constant, u is wind speed at the reference height z, d is the zero plane displacement, and z_0 is the roughness height. These parameters were expressed using the tree height h; $d = 0.78h$, and $z_0 = 0.07h$ (Hattori, 1985). Measured wind speed was used assuming $z = h + 1$ m, neutral stability and the same conductance for heat and momentum. As relative humidity measurable by a high-polymer hygrometer at the meteorological station is less than 95%, hourly average relative humidity during rainfall at the Mito Meteorological Observatory (25 km southwest of Hitachi Ohta, the hygrometer is Vaisala HMP233), Japan Meteorological Agency, was employed on a monthly basis if relative humidity is over 95%. The accuracy of Vaisala HMP223 was within ±2% for 90% ≤ relative humidity ≤ 100%, and ±1% for 0% ≤ relative humidity ≤ 90%. Eqs. (5) to (8) are solved numerically as T_s, H, and E are unknown. Using obtained value of E, I is calculated with Eqs. (1) and (2). You need to note that the heat budget approach assumes evaporation from canopy surface only, while the SDE model in the next section includes evaporation from both splash droplets and canopy surface.

SDE Model

As shown in the section THE IDEA OF SDE, hourly canopy interception is proportional to hourly rainfall (rainfall intensity), and an inclination of the regression line is defined as the DOCIORI, i. The SDE model is constructed empirically based on this trait. Hourly evaporation rate E is a product of i and hourly rainfall p_g,

$$E = ip_g \qquad\qquad (10)$$

In Figure 5 DOCIORI is potted against DOY (day of year) for two years of observation 1999 to 2000, and reveals seasonal changes with a peak in summer. Figure 5 represents that evaporation by canopy interception can be expressed as a function of DOY, and the regression curve is presented. In winter when no data were available constant values of DOCIORI (0.1) were assumed, since minimum figures of DOCIORI scatter around some 0.1. The function is expressed as,

$$i = 0.243 \sin\left(\frac{DOY}{2.16}\right) - 0.0184 \qquad \text{for } 63 < DOY < 326 \qquad (11)$$

$$i = 0.1 \qquad\qquad \text{for } DOY \leq 63 \text{ or } DOY \geq 326 \qquad (12)$$

Unit of the sine function is degrees. Canopy interception I for the SDE model is obtained by combining Eqs. (10) – (12) and Eqs. (1) and (2). As the SDE model includes splash droplet evaporation as well as evaporation from canopy surface, the output is expected to be different from that of the heat budget model that considers evaporation from canopy surface only; the difference between the two models can be regarded as the amount of SDE (see DISCUSSION).

Revised Gash Model

The revised Gash model (Gash, 1995) is based on the original Gash model (Gash, 1979). The former is applicable to the sparse forest stand, while the latter is not in some cases, though the principle structure of the both versions is the same. In this respect the revised Gash model (referred to as the Gash model hereafter) has wider applicability and is adopted in this chapter. The Gash model is practically calculated from daily rainfall totals assuming one storm per rainday, however, in this chapter it is calculated from a rainfall total of a discrete rain event setting a separation time of 6 hours (see the section of SITE AND INSTRUMENTATION). At first the model was applied monthly, but the analysis failed since the data included some months in which the number of rain events was too small to run the model. Finally, the application was made bimonthly and annually.

The Gash model considers three stages of evaporation for a discrete rain event that is similar to the backbone model. Firstly, a period of wetting up when P_G is less than the threshold value necessary to saturate the canopy, P'_G. Secondly, a period of saturation, and thirdly, a period of drying out after the end of the rainfall. P'_G is written as,

$$P'_G = -\frac{\overline{R}S_c}{\overline{E}_c} \ln\left[1 - \left(\frac{\overline{E}_c}{\overline{R}}\right)\right] \tag{13}$$

The bars on R and E_c represent the average over the time t from canopy saturation to the end of the rain event. $\overline{R}$ is expressed as $\overline{R} = \frac{P_G - P'_G}{t}$, which is practically equal to average p_g for $P_G \gg P'_G$. Canopy capacity and mean evaporation per unit area of cover, S_c and $\overline{E}_c$, are written as $S_c = S/c$ and $\overline{E}_c = \overline{E}/c$, respectively: the subscript "c" represents per unit area of cover. The value of S_c was 0.51 mm in 1999 and 0.47 mm in 2000. $\overline{E}$ is mean evaporation over the time t, and is expressed as a product of $\overline{R}$ and the coefficient of a in Eq. (14) that is a inclination of a regression line between rainfall and canopy interception. The figures of τ and t are largely equal for a large rain event. $\overline{E}$ and E are also almost equal if the equivalent time scales are considered.

Evaporation processes are categorized by the value of P'_G and the rainfall scheme as follows,

1)   Evaporation for m number of storms that do not saturate the canopy ($P_G < P'_G$)
2)   Evaporation for n number of storms that wet-up the canopy to saturation ($P_G \geq P'_G$)
3)   Evaporation from saturated canopy during rainfall
4)   Evaporation once rainfall ceases

Once the canopy is saturated evaporation from the trunks are also taken into consideration that consists of two processes depending on the amount of rainfall.

1)   Evaporation for q number of storms that saturate tree trunks ($P_G \geq S_t/p_t$)
2)   Evaporation for n–q number of storms that do not saturate the tree trunks ($P_G < S_t/p_t$)

The values of p_t, the proportion of rainwater diverted to stemflow, were 0.042 in 1999 and 0.050 in 2000.

INTERRELATION OF THE MODELS

Though all the models described in the previous section except the evaporation component of SDE model are physically based, the heat budget model does not fit well the observed values in some cases while the Gash models works well as mentioned in INTRODUCTION (specific examples are shown in RESULT and DISCUSSION). The insufficiency of the heat budget model is attributable to the overlook of SDE, but the Gash models include SDE implicitly. In this section it is shown that the heat budget model contains contradiction due to the oversight of SDE and that the SDE and the Gash models are mathematically equivalent.

Basic Equation Between the Backbone Model and the Gash Models

The backbone model and the Gash models include similar variables, E and $\overline{E}$, and τ and t. As shown in the previous section E and $\overline{E}$, and τ and t are largely equal, respectively. The aim of this section is to introduce Eq. (16) that is an important equation to understand the relation between the backbone model and the Gash model (including the original Gash model).

Canopy interception can be expressed using an empirical equitation,

$$I = aP_G + b \qquad (14)$$

where a and b are regression coefficients. Eq. (14) is valid for series of discrete rain data regarding I and P_G as variables. Comparing Eq. (2) and (14) E is expressed as,

$$E = a\left(\frac{P_G}{\tau}\right) \qquad (15)$$

Considering $P_G \approx P_G - P'_G$ and $\tau \approx t$ for large P_G, $P_G / \tau \approx P_G - P'_G / t = \overline{R}$. Since the coefficients a and b are mainly determined by large P_G, this approximation is valid. As a result,

$$\overline{E} = a\overline{R} \qquad (16)$$

Eq. (16) is derived from the backbone model using approximation, but the same equation is described in Gash (1979) that introduces the original Gash model. This implies that Eq. (16) holds true for the heat budget model, the Gash models, and the SDE model.

Contradiction of the Heat Budget Model

Eq. (16) claims that the mean evaporation rate is proportional to the mean rainfall intensity with the proportional coefficient a. On the other hand, the heat budget model does not include rainfall intensity as a variable in Eqs. (5) – (8) and contradicts Eq. (16). The physical implication in Eq. (16) is understandable by introducing the idea of SDE, however, the heat budget model does not involve the idea. Specific examples for the contradiction are shown in RESULT and DISCUSSION.

Forest hydrologists have not understood a simple Eq. (16) to the letter, but have regarded it as one of the mathematical processes to derive the original Gash model. Nothing is too simple to be misunderstood; however, there are some reasons that the forest hydrologists overlooked the physical meaning of Eq. (16) and should not be blamed.

From the view point of the modeling, most canopy interception models are constructed on a rain event basis considering interception processes; wetting up, saturation, and drying out. For this reason, most canopy interception models are applied on a rain event or a daily basis, even if higher time resolution of the data that can detect DOCIORI (e.g. hourly data, see DISCUSSION) were obtained. Taking an average over a rain event for $\overline{R}$ and $\overline{E}$ degrades DOCIORI. As for the Gash models, $\overline{R}$ and $\overline{E}$ are calculated as an average of longer period, e.g. biweekly, monthly, or a few month, that makes the situation worse. On the other hand, in many canopy interception sites where the rainfall intensity is low, the heat budget models worked well because the contribution of the SDE seemed minor.

Equivalency of the SDE Model and the Gash Models

As described in the previous section, the heat budget model overlooked SED but the Gash models include it. In this section it is shown that the Gash models (both the original and revised version) includes SDE implicitly and that the SDE model and the Gash models are mathematically equivalent.

The following equation is derived from the definition of the SDE model, Eq. (10),

$$E\tau = \Sigma(ip_g) \tag{17}$$

The summation on the right side of Eq. (17) is made for hours of rainfall. Considering i is independent from p_g and $\Sigma p_g = P_G$, Eq. (17) is written as,

$$E\tau = iP_G \tag{18}$$

The following equation is obtained combining Eq. (18) and Eq. (2).

$$I = iP_G + S_a \tag{19}$$

If we regard $i = a$ and $S_a = b$ Eq. (19) has the same form with Eq. (14) that is the basic equation to derive the Gash models. Now, it is proved that the SDE model and the Gash models are mathematically equivalent. Forest hydrologists have applied the Gash models without notice that the SDE process is built-in in them.

RESULT

Snowfall, occult precipitation, and negative interception values due to error were excluded from the analysis. As a result, annual rainfall amount for the analysis was 1673.3 mm in 1999, and 1365.0 mm in 2000 (Table 3), while total observed precipitation was 1707.9 mm in 1999, and 1452.9 mm in 2000 (Table 1). Annual interception was 320.1 mm in 1999 (interception rate 19.1%), and 256.4 mm in 2000 (18.9%).

Table 3. Observed rainfall, observed canopy interception and calculated components of canopy interception by the revised Gash model

Year	1999	2000
Observed rainfall (mm)	1673.3	1365.0
Observed canopy interception (mm)	320.1	256.4
Evaporation for small storms that are insufficient to saturate the canopy (%)	2.0	2.4
Evaporation for storms that wetting-up the canopy to saturate (%)	1.1	1.3
Evaporation from the saturated canopy during rainfall (%)	80.1	73.9
Evaporation after the rainfall ceases (%)	8.9	11.9
Evaporation from the trunks (%)	7.9	10.4

The sum of the components in 2000 is not 100% due to the rounding error.
(Revised and reproduced from Murakami, 2007a; Copyright Elsevier B.V.)

Estimated canopy interception with the heat budget model was 118.7 mm in 1999 and 93.9 mm in 2000, respectively. These values are only one third of measured values (Figure 6 (a), (b)). Calculated values do not rise with observed values, and plateau at around 10 mm. The values of r_a were $18.6/u$ s m^{-1} in 1999 and $17.8/u$ s m^{-1} in 2000, respectively, for the tree height of 5.8 m and 6.3 m. Predicted interception using the SDE model well reproduced observed values on an annual basis and on a rain event basis as well (Figure 6 (c), (d)). Estimated and observed values were 353.5 mm and 320.1 mm in 1999 and 294.9 mm and 256.4 mm in 2000, respectively.

The Gash model was applied annually and bimonthly by changing the parameters (a, $\overline{R}$, $\overline{E}$, $\overline{E}_c$, P'_G) on an annual and a bimonthly basis, respectively, while forest parameters (c, S, S_t, p_t) were set at constant through each year for both the application runs. Estimated interception by the annually applied Gash model was 349.4 mm (observation 320.1 mm) in 1999 and 319.0 mm (observation 256.4 mm) in 2000, respectively (Murakami, 2007b). While the bimonthly applied Gash model showed better annual estimation; 323.1 mm in 1999 and 291.2 mm in 2000 (Figure 6 (e), (f)). The application of the Gash model in a shorter time interval is superior to in a longer interval since the former can reflect seasonal changes in interception but the latter cannot. Bimonthly estimated interception, i.e. each data point in Figure 6 (e), (f), fit well with observed values. Annual evaporation components estimated by the bimonthly applied Gash model are listed in Table 3 that shows evaporation during rainfall is a major component of interception at the site; 80.1% and 73.9% of the total interception in 1999 and 2000, respectively.

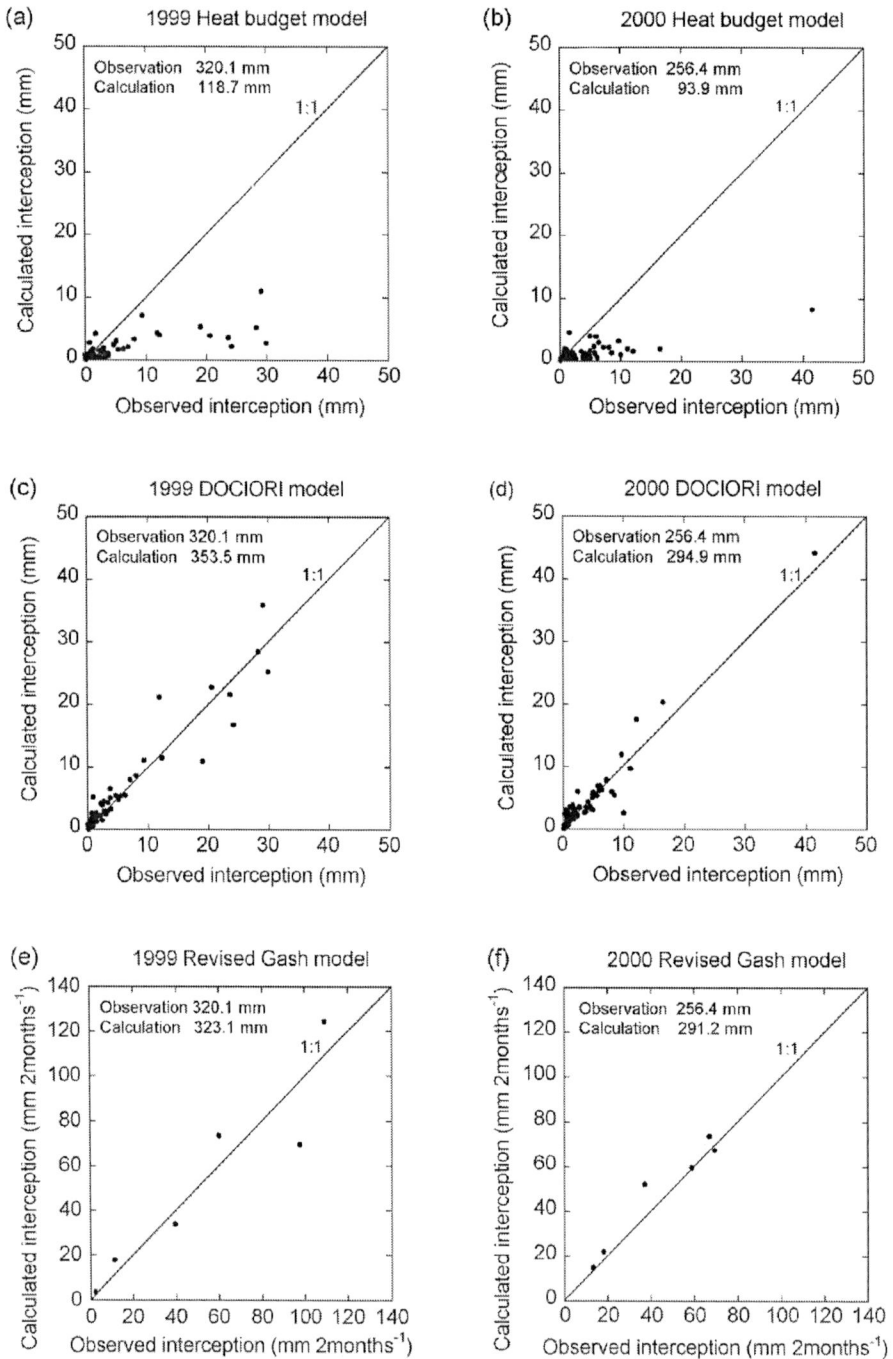

Figure 6. Measured and calculated canopy interception using three interception models. (a) Heat budget model for 1999. (b) Heat budget model for 2000. (c) SDE (DOCIORI) model for 1999. (d) SED (DOCIORI) model for 2000. (e) Revised Gash model for 1999. (f) Revised Gash model for 2000. (a) – (d) were complied on a rain event basis, but (e) and (f) were on bimonthly basis.

DISCUSSION

Reconsideration of the Heat Budget Model

Underestimation of calculated canopy interception by the heat budget model (Figure 6 (a), (b)) could be attributable to inappropriate forest and/or hydrometeorological parameters. To evaluate the possibility the sensitivity test was conducted by changing relative humidity, $r_a u$, and S_{max}. First of all, relative humidity was reduced by 5% that is equivalent to the accuracy of hygrometer at the site but larger than that at the Mito Meteorological Observatory described in the section of *Heat budget model* in THREE CANOPY INTERCEPTION MODELS. The calculated annual interceptions were 149.6 mm in 1999 and 119.8 mm in 2000, respectively, and were still less than half of the measured values of 320.1 mm in 1999 and 256.4 mm in 2000, respectively. Secondly, values of $r_a u$ were diminished by half from 18.6 to 9.3 in 1999 and from 17.8 to 8.9 in 2000, respectively. Assuming $u = 1$ m s^{-1} the equivalent tree heights were 58 m and 92 m that are unrealistically high. Calculated interceptions were 175.6 mm in 1999 and 131.3 mm in 2000, which were around half of the measured values. Thirdly, S_{max} was doubled, and the results were 153.9 mm in 1999 and 130.9 mm in 2000 that were less than half of the observed values again. Though relative humidity, $r_a u$, and S_{max} were changed individually, it was difficult to explain the discrepancy between calculated and measured values by the error of these variables even if they varied in tandem.

Some forest hydrologists noticed that heat budget approaches underestimated canopy interception, and dealt with or understood the results in different ways. Calder et al. (1986) measured and calculated canopy interception in West Java. They reduced r_a (5 s m^{-1}) and enhanced S (4.5 mm) for optimization of the estimated results. Schellekens et al. (1999) applied the Rutter model (Rutter et al. 1971; 1975) that contains the heat budget component and the original Gash model to a lowland rain forest in Puerto Rico. The Rutter model underestimated canopy interception for large rain events with estimated $\overline{E}$ of 0.11 mm h^{-1} and S of 1.15 mm. They optimized the results changing $\overline{E}$ and S. Larger $\overline{E}$ (2.80 mm h^{-1}) showed better fit than larger S (5.75 mm) or larger $\overline{E}$ (1.10 mm h^{-1}) and S (5.75 mm). They also state that very large storage capacities for tropical rain forest (Herwitz, 1985; Calder et al, 1986) are not supported by their analysis. Iida et al. (2005) observed canopy interception in a stand of Japanese red pine and applied the Penman-Monteith equation that is a sort of heat budget models to the stand. They estimated rainwater storage on the tree surface as the difference between measured canopy interception and calculated evaporation. Their results show that the storage capacity increases with rainfall amount. The maximum rainwater storage was 6.5 mm with a rainfall amount of 50 mm. They state this value was reasonable referring to Herwitz (1985), but it is unusual that the water storage in the canopy increases with rainfall of up to 50 mm. Since rainfall intensity increases with rainfall amount (Kondo et al., 1992), evaporation also augments with the rainfall amount due to SDE. The use of the Penman-Monteith equation that does not consider SDE underestimates evaporation during rainfall, and the residual was imposed on the rainwater storage in the canopy. Vernimmen et al. (2007) stated that the average rate of $\overline{E}$ calculated by the Penman-Monteith equation was only 0.06 mm h^{-1} while the figure estimated from water budget was 1.4 mm h^{-1} for stunted Heath Forest in Central Kalimantan, Indonesia.

All the studies mentioned in this section underestimate canopy interception, and the sites were tropical areas or monsoonal Asia where high intensity rainfalls tend to occur. In such areas SDE or DOCIORI was observed without being noticed by the scientists. It is shown that this is indeed the case in the section SDE other than Japanese sites in THE IDEA OF SDE.

Amount of SDE

As described in the section *SDE model* in THREE CANOPY INTERCEPTION MODELS, the amount of SDE can be obtained as the difference between observed and estimated canopy interception by the heat budget model, since the heat budget model estimates evaporation only from canopy surface. Referring to Figure 6 (a) and (b), the values of SDE are 201.4 mm (= 320.1 − 118.7 mm; 62.9% of observed interception) in 1999 and 162.5 mm (= 256.4 − 93.9 mm; 63.4% of observed interception) in 2000. These values are consistent with the predicted figures by the Gash model. The maximum amount of SDE can be estimated as the sum of the three components shown in Table 3; evaporation for small storms that are insufficient to saturate the canopy, evaporation for storms that wetting-up the canopy to saturate, and evaporation from the saturated canopy during rainfall yielding 83.2% in 1999 and 77.6% in 2000.

Toba et al. (2006) estimated SDE employing the same approach with the above-mentioned; combination of the water budget and the heat budget model. The experiment procedures were the same with Toba and Ohta (2008). They used 24 vinyl artificial trees with 60 cm high and a rainfall simulator that were installed on the outdoor condition, and conducted micrometeorological observation above canopy. The amount of SDE (they termed it "mist") M, was calculated from observed rainfall P_G, observed net rainfall TS, evaporation from canopy surface $E\tau$ estimated by the Penman-Monteith equation (heat budget model), and water storage on the canopy 10 minutes following the cessation of the rainfall d.

$$M = P_G - TS - E\tau - d \qquad\qquad\qquad (20)$$

P_G was measured by a tipping bucket rain gauge, TS was collected by a tray and measured using a tipping bucket, and d was weighed by an electric balance. They concluded that the amount of M was around 60% of total interception that is comparable with the estimation above.

Nakayoshi et al. (2007) observed rainfall interception at an outdoor urban scale model site where 1.5-m cubic roughness blocks were arranged in a 50 m × 100 m area. Rainfall interception was measured in the 6 m × 6 m area using water balance where the entire surface was sealed to avoid leakage and absorption. Rainfall intensity ranged from 0.6 − 1.4 mm h^{-1} on average with the maximum of 4.6 mm h^{-1}. They could not detect the DOCIORI, and concluded that the evaporation during rainfall is explainable by surface evaporation calculated from the micrometeorological data including ground heat flux. In urban canopy no space exists under the surface of the canopy, while forest canopy includes much space under the canopy surface that might promotes SDE.

A Clue to Solve Enormous Evaporation

Why Does the Laundry Dry on a Rainy Day?

It is an observational fact that enormous evaporation occurs from the forest canopy during rain events, regardless of the evaporation mechanisms—whether it is SDE or surface evaporation. Though evaporation is dependent on some hydrometeorological variables, air temperature, wind speed, net radiation, and water vapor pressure deficit, these variables are not necessarily the causes of evaporation but in some cases some of them are determined as a result of heat and water vapor transport. You cannot understand the mechanism of canopy interception by the measurement of the variables only. Specifically, water vapor should be removed to keep the air unsaturated, otherwise water vapor reaches saturation and evaporation stops. This problem is beyond a matter of canopy interception; it is a meteorological problem.

You can dry the laundry on a rainy day when the ground surface gets drenched and numerous rain drops are in the air, which indicates water vapor does not saturate during rainfall. As mentioned in the section INTRODUCTION evaporation rate of 10 mm h^{-1} requires latent heat of 6814 W m^{-2}, five times the solar constant. How and from where such enormous energy is supplied is not known, and it is amazing that this contradiction has not attracted scientists' attention. One reason is the difficulty in the measurement of meteorological variables during rainfall, and another is lack of theory. Concerning water vapor transfer, an innovative theory that is a clue to solving the problem was proposed by Makarieva and Gorshkov (2007); it is described in the following sections.

Difference in Precipitation Between Forested and Non-Forested Landmass

A body of theory that may elucidate the huge evaporation during rainfall was originally constructed to explain precipitation distribution on a continental scale. Makarieva and Gorshkov (2007) reviewed the relation between the distance from the ocean measured along the streamline, x, and annual precipitation, P, for eight terrestrial transects of the International Geosphere Biosphere Program (Figure 7). Physically thinking, precipitation decreases with x, because some of the precipitated water is lost as river water and groundwater that flow back into the ocean and the residual is transported to the inner part of the continents. Mathematically, P diminishes exponentially with x, and the data in Figure 8 (a), five non-forested transects, clearly demonstrate this trend. On the contrary, for three forested transects the Amazon, Congo, and Yenisey river basin, P remains constant or increases with x that violates the physical principle (Figure 8 (b)). This implies that the forest works actively to transport water vapor inland as "the biotic pump of atmospheric moisture". Makarieva and Gorshkov (2007) formulated the physical principle that dictates biotic pump, termed "evaporative force", which is summarized in the next section.

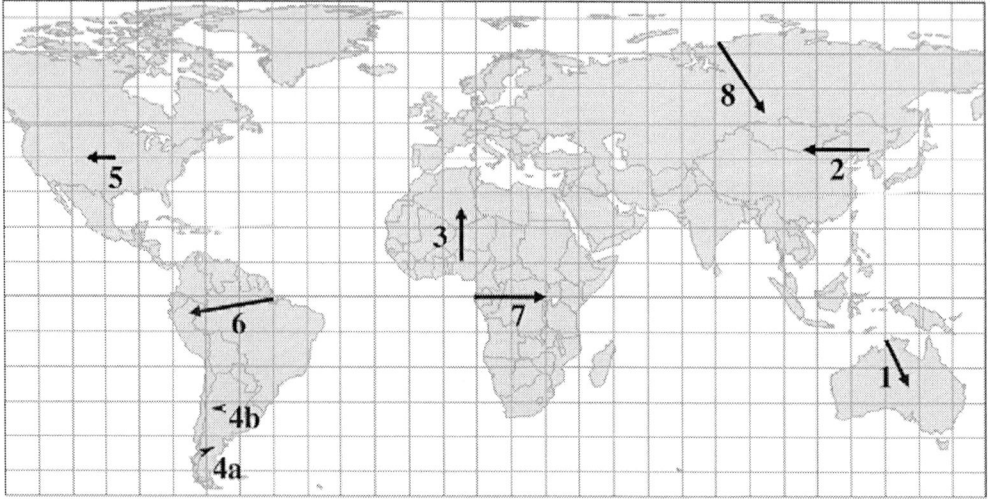

Figure 7. Transacts where dependence of annual precipitation P on the distance from the ocean x were analyzed.

Evaporative Force

In this section evaporative force is outlined qualitatively. For mathematical formulation you have to refer to the original paper, Makarieva and Gorshkov (2007). The following three physical properties of water vapor produce "evaporative force".

1) Saturated water vapor pressure decreases with height, and excessive water vapor condenses and removed as cloud or precipitation
2) Molecular weight of H_2O is 18 that is smaller than that of dry air, 29
3) The lapse rate of saturated water vapor, Γ_{H2O}, is 1.2 K km^{-1}, much smaller than that of dry or moist adiabatic lapse rate, Γ_a, 9.8 or ≈ 6 K km^{-1}, respectively.

Dry air is aerostatic equilibrium (left in Figure 9), however, reflecting above physical properties, atmospheric water vapor is not aerostatic equilibrium. Assuming saturated atmospheric water vapor, and remove dry air from the air column for simplicity. Partial pressure of water vapor is not compensated by the weight of water vapor in the air column (right in Figure 9), and this yields upward directed evaporative force.

Conventional atmospheric motion considers stability-instability condition for an air parcel based on dry or moist adiabatic lapse rate, Γ_a. If the lapse rate of the ambient air Γ is steeper than that of the air parcel Γ_a, $\Gamma > \Gamma_a$, incidental small amount of upward displacement of the air parcel, i.e. partial heating, triggers convection. The air parcel continues to rise due to buoyancy up to the tropopause if this condition is maintained, though it is difficult to estimate the motion of the air mass that induced by the convection quantitatively. Inversely, for $\Gamma < \Gamma_a$, the air parcel is stable due to lack of buoyancy.

Contrary to the conventional idea, evaporative force always works without partial heating if $\Gamma > \Gamma_{H2O}$, and can be dealt with quantitatively.

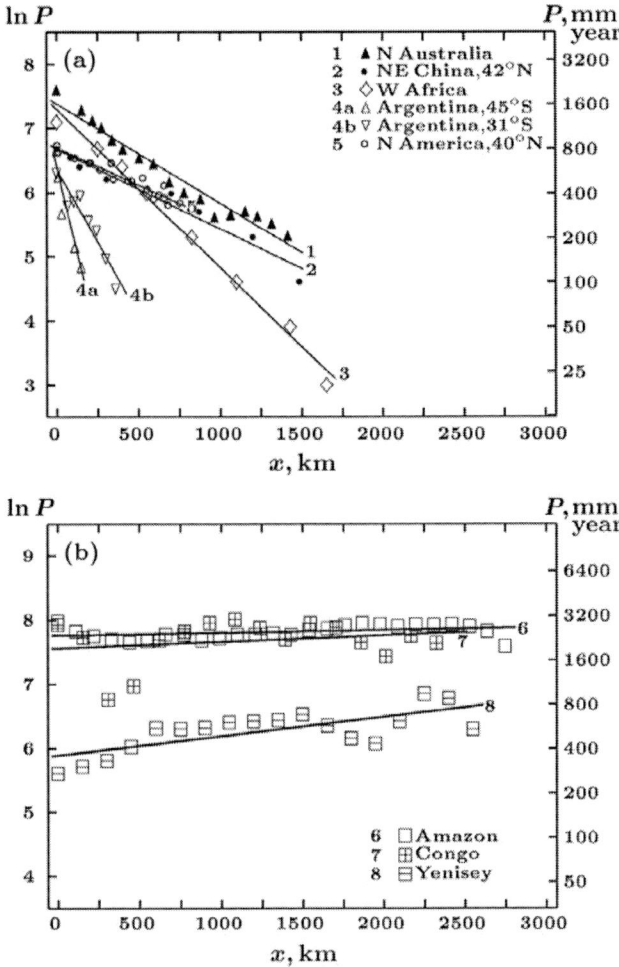

(Reproduced by permission from Makarieva and Gorshkov, 2007; Copyright Makarieva and Gorshkov).

Figure 8. Dependence of precipitation P on the distance from the ocean x. (a) non-forested territories, and (b) forested territories.

For saturated water vapor with the ground temperature of 15°C evaporative force that decreases with an altitude is calculated as some 7 hPa km^{-1} at the sea surface level and around 3 hPa km^{-1} at the altitude of 2 km (Figure 3 (b) in Makarieva and Gorshkov, 2007). Evaporative force gets stronger with moisture and the surface with intensive evaporation "sucks in" moist air from the surface with weak evaporation; forest evaporates much water than the ocean and forest sucks in water vapor from the ocean as horizontal flow. Precipitation in Yenisey river basin increases with decreasing latitude, i.e. increasing x, (Figure 8 (b)) that is explainable by evaporative force; saturated water vapor pressure is higher at lower latitude due to higher temperature that boosts evaporative force at lower latitude. Evaporative force can explain not only biotic pump but also some atmospheric motions that were not satisfactorily understood. The General Circulation Models combined with evaporative force would change the present estimation on the effect of deforestation drastically. The existence of desert, monsoon, and trade wind are also explainable by evaporative force.

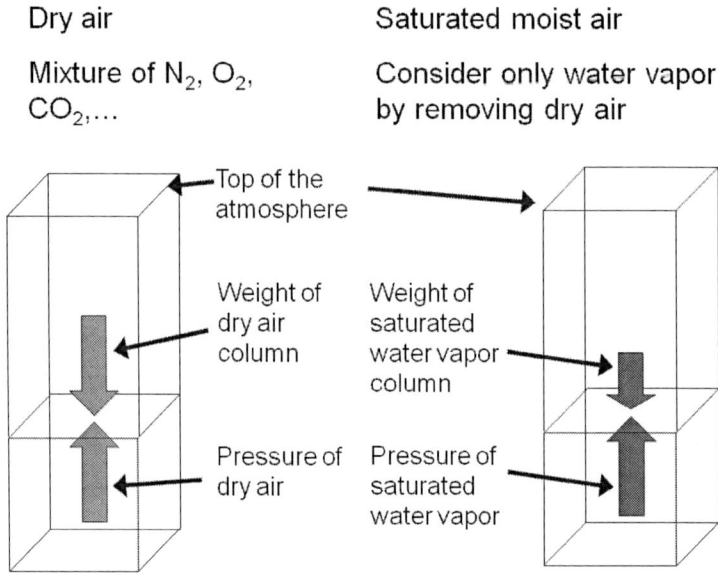

Figure 9. Conceptual explanation for evaporative force. *Left*: the ratio of gas components of dry air is constant with respect to height (Dalton's law), and dry air is under aerostatic equilibrium. *Right*: consider saturated moist air, and remove dry air from it for simplicity. Partial pressure of water vapor is not compensated by the weight of water vapor in the air column, and this difference causes evaporative force.

Evaporative force is powered by high humidity during rainfall that greatly promotes evaporation. Efficient transport of water vapor during rainfall is fully understandable by this process, and a huge amount of water vapor is sucked in from the ocean to compensate for canopy interception. In spite of the importance of canopy interception from the view point of water budget, Makarieva and Gorshkov (2007) describe canopy interception only in one sentence, and they mainly focus on evaporation other than canopy interception.

Latent heat transfer and the source of latent heat relevant to canopy interception are also explainable in connection with water vapor transfer as follows. During rainfall, raindrops fall from the cloud down to the ground, and raindrops pull down ambient air with latent heat released by condensation in the cloud. The raindrops and accompanied air that flow down push out the air near the surface upward, which is cooled by evaporation of canopy interception. Latent heat consumed by canopy interception and released by condensation in the cloud are exchanged by mixing, which makes ends meet. This assumption about heat and water vapor exchange is consistent, but the proof seems to be more difficult than the proof of evaporative force itself whether it is theoretical or observational, because the phenomenon occurs under the rainfall condition.

CONCLUSION

The phenomenon of DOCIORI was noticed in the 1980s by Japanese forest hydrologists, but the mechanism remained unsolved for a few decades. The idea of SDE can elucidate the cause of the DOCIORI that is also useful to resolve the contradiction on the heat budget model. The heat budget model underestimates observed interception in some sites where rainfall intensity is high, because it considers evaporation only from the canopy surface overlooking SDE. Contrary to the heat budget model, generally speaking, widely used Gash models well reproduce measured interception, because they include SDE implicitly. The idea of SDE solved some problems relevant to canopy interception, while it does not come to a happy end. An enormous amount of evaporation is caused by efficient water vapor transport mechanism and supply of equivalent latent heat. A clue to solve this enigmatic problem lies in the evaporative force that is an unnoticed basic principle in meteorology. Conversely, proof of evaporative force may be made via canopy interception that implies canopy interception science would contribute to establishing the basic principle of atmospheric sciences. SDE and evaporative force have not proved directly yet, and are waiting for more observational evidence and theoretical development.

REFERENCES

Bruijnzeel, L. A., Wiersum, K. F., 1987. Rainfall interception by a young Acacia auriculiformis (a. cunn) plantation forest in West Java, Indonesia: Application of Gash's analytical model. *Hydrological Processes.* **1**, 309-319.

Calder, I.R., Wright, I.R., Murdiyarso, D., 1986. A study of evaporation from tropical rain forest-West Java. *Journal of Hydrology.* **89**, 13-31.

Deguchi, A., Hattori, S., Park, H., 2006. The influence of seasonal changes in canopy structure on interception loss: Application of the revised Gash model. *Journal of Hydrology.* **319**, 80-102.

Dykes, A. P., 1997. Rainfall interception from lowland tropical rainforest in Brunei. *Journal of Hydrology.* **200**, 260-279.

Gash, J. H. C., 1979. An analytical model of rainfall interception by forests. *Quarterly Journal of the Royal Meteorological Society.* **105**, 43-55.

Gash, J. H. C., Lloyed, C. R., Lachaud, G., 1995. Estimating sparse forest rainfall interception with an analytical model. *Journal of Hydrology.* **170**, 79-86.

Hashino, M., Tamura, T., 2005. Ranryu riron ni yoru jyukansyadan ni shimeru bisaisuiteki no bunri hyouka (Separation and evaluation of small droplets making up canopy interception using turburant theroy). *Proceding of Annual Conference Japan Society of Hydrology and Water Resou*rces. 260-261 (in Japanese). http://www.jstage.jst.go.jp/article/jshwr/18/0/18_128/_article

Hattori, S., 1985. Explanation on derivation process of equations to estimate evapotranspiration and problems on the application to forest stand. *Bulletin of the Forestry and Forest Products Research Institute.* **332**, 139-165 (in Japanese with English summary).

Hattori, S., Chhikaarashi, H., Takeuchi, N., 1982. Measurement of the rainfall interception and its micrometeorological analysis in a Hinoki stand. *Bulletin of the Forestry and Forest Products Research Institute.* **318**, 79-102 (in Japanese with English summary).

Herwitz, S.R., 1985. Interception storage capacities of tropical rainforest canopy trees. *Journal of Hydrology.* **77**, 237-252.

Hörman, G., Branding, A., Clemen, T., Herbst, M., Hinrichs, A., 1996. Calculation and simulation of wind controlled canopy interception of beech forest in Northern Germany. *Agricultural and Forest Meteorology.* **79**, 131-148.

Iida, S., Tanaka, T., Sugita, M., 2005. Change of interception process due to the succession from Japanese red pine to evergreen oak. *Journal of Hydrology.* **315**, 154-166.

Kondo, J., Nakamura, T., Yamazaki, T., 1991. Estimation of the solar and downward atmospheric radiation. *Tenki (the Bulletin Journal of the Meteorological Society of Japan).* **38**, 41-48 (in Japanese).

Kondo, J., Watanabe, T., Nakazono, M., Ishii, M., 1992. Estimatin of forest rainfall interception. *Tenki (the Bulletin Journal of the Meteorological Society of Japan).* **39**, 159-167 (in Japanese).

Leyton, L., Reynolds, E. R. C., Thompson, F. B., 1967. Rainfall interception in forest and moorland. In: *International Symposium on Forest Hydrology*, Sopper, W. E., Lull, H. W. , Ed.; Pergamon Press, Oxford, 163-178.

Link, T. E., Unsworth, M., Marks, D., 2004. The dynamics of rainfall interception by a seasonal temperate rainforest. *Agricultural and Forest Meteorology.* **124**, 171-191.

Llorens P., Poch, R., Latron, J., Gallart, F., 1997. Rainfall interception by a *Pinus sylvestris* forest patch overgrown in Mediterranean mountainous abandoned area I. Monitoring design and results down to the event scale. *Journal of Hydrology.* **199**, 331-345.

Makarieva, M. A., Gorshkov, V. G., 2007. Biotic pump of atmospheric moisture as driver of the hydrological cycle on land. *Hydrology and Earth System Sciences.* **11**, 1013-1033.

Marshall, J. S., Palmer, W.M., 1948. The distribution of raindrops with size. *Journal of Meteorology.* **5**, 165-166.

Murakami, S., 2006. A proposal for a new forest canopy interception mechanism: Splash droplet evaporation. *Journal of Hydrology.* **319**, 72-82.

Murakami, S., 2007a. Application of three canopy interception models to a young stand of Japanese cypress and interpretation in terms of interception mechanism. *Journal of Hydrology.* **342**, 305-319.

Murakami, S., 2007b. Analysis of canopy interception in a Japanese cypress stand using Gash models. *Transactions of Kanto Branch of the Japanese Forest Society.* **58**, 149-151 (in Japanese).

Murakami, S., 2008. Unveiled Evaporation Mechanism of Forest Canopy Interception. In: *New Topics in Water Resources Research and Management*; Andreassen, H. M.; Ed.; Nova Science Publishers, New York, 279-296.

Murakami, S., Tsuboyama, Y., Shimizu, T., Fujieda, M., Noguchi, S., 2000. Variation of evapotranspiration with stand age and climate in a small Japanese forested catchment. *Journal of Hydrology.* **227**, 114-127.

Nakayoshi, M., Moriwaki, R., Kanda, M., 2007. Preliminary results of rainfall interception over the outdoor urban scale model. *Annual Journal of Hydraulic Engineering.* **51**, 247-252 (in Japanese with English abstract).

Pearce, A. J., Gash, J. H. C., Stewart, J.B., 1980. Rainfall interception in a forest stand estimated from grassland meteorological data. *Journal of Hydrology.* **46**, 147-163.

Pyker, T. G., Bond, B. J., Link, T. E., Marks, D., Unsworth, M. H., 2005. The importance of canopy structure in controlling the interception loss of rainfall: Examples from a young and an old-growth Douglas-fir forest. *Agricultural and Forest Meteorology.* **130**, 113-129.

Rutter, A. J., Kershaw, K. A., Robins, P. C., Morton, A. J., 1971. A predictive model of rainfall interception in forests I. Derivation of the model from observations in a plantation of Corsican pine. *Agricultural Meteorology.* **9**, 367-384.

Rutter, A. J., Morton, A. J., Robins P. C., 1975. A predictive model of rainfall interception in forests II. Generalization of the model and comparison with observations in some coniferous and hardwood stands. *Journal of Applied Ecology.* **12**, 367-380.

Schellekens, J., Scatena, F.N., Bruijnzeel, L. A., Wickel, A. J., 1999. Modelling rainfall interception by a lowland tropical rain forest in northern Puerto Rico. *Journal of Hydrology.* **225**, 168-184.

Sidle, R. C., Tsuboyama, Y., Noguchi, S., Hosoda, I., Fujieda, M., Shimizu, T., 1995. Seasonal hydrologic response at various spatial scales in a small forested catchment, Hitachi Ohta, Japan. *Journal of Hydrology.* **168**, 227-250.

Sidle, R. C., Tsuboyama, Y., Noguchi, S., Hosoda, I., Fujieda, M., Shimizu, T., 2000. Stormflow generation in steep forested headwaters: a linked hydrogeomorphic paradigm. *Hydrological Processes.* **14**, 369-385.

Singh, B., Szeicz, G., 1979. The Effect of Intercepted Rainfall on the Water Balance of a Hardwood Forest. *Water Resources Research.* **15**, 131-138.

Stewart, J. B. 1977. Evaporation from the wet canopy of a pine forest. *Water Resources Research.* **13**, 915-921.

Toba, T., Ohta, T., Shiraki, K., Enatsu, T., 2006. Characteristics of rainfall interception loss using tree models. Consideration of the interception loss in rainfall. *The Japanese Forest Society Congress.* **117**, 459 (in Japanese). http://www.jstage.jst.go.jp/article/jfsc/117/0/117_459/_article

Toba, T., Ohta, T., 2008. Factors affecting rainfall interception determined by a forest simulator and numerical model. *Hydrological Processes.* **22**, 2634-2643.

Tsuboyama, Y., Sidle, R. C., Noguchi, S., Hosoda, I., 1994. Flow and solute transport through the soil matrix and macropores of a hillslope segment. *Water Resources Research.* **30**, 879-890.

Tsuboyama, Y., Sidle, R. C., Noguchi, S., Murakami, S., Shimizu, T., 2000. A zero-order basin - its contribution to catchment hydrology and internal hydrological processes. *Hydrological Processes.* **14**, 387-401.

Tsukamoto, Y., Tange, I., Minemura, T., 1988. Interception loss from forest canopies. *Hakyuchi-kenkyu (Bulletin of the Institute for Agricultural Research on Rolling Land).* **6**, 60-82 (in Japanese with English summary).

van der Tol, C., Gash, J. H. C., Grant, S. J., McNeil, D. D., Robinson, M., 2003. Average wet canopy evaporation for a Sitka spruce forest derived using the eddy correlation-energy balance technique. *Journal of Hydrology.* **276**, 12-19.

Vernimmen, R. R. E., Bruijnzeel, L. A., Romdoni, A., Proctor, J., 2007. Rainfall interception in three contrasting lowland rain forest types in Central Kalimantan, Indonesia. *Journal of Hydrology.* **340**, 217-232.

Wallace, J., McJannet, D., 2006. On interception modelling of a lowland coastal rainforest in northern Queensland, Australia. *Journal of Hydrology.* **329**, 477-488.

Wallace, J., McJannet, D., 2008. Modelling interception in coastal and montane rainforests in northern Queensland, Australia. *Journal of Hydrology.* **348**, 480-495.

Waterloo, M. J., Bruijnzeel L. A., Vugts, H. F., 1999. Evaporation from Pinus Caribaea Plantations on Former Grassland Soils Under Maritime Tropical Condit. *Water Resources Research.* **35**, 2133-2144.

In: Forest Canopies: Forest Production, Ecosystem…   ISBN 978-1-60741-457-5
Editor: J. D. Creighton and P. J. Roney   © 2009 Nova Science Publishers, Inc.

Chapter 2

EXOTIC HERB LAYERS AS ECOLOGICAL FILTERS IN FOREST UNDERSTORIES

*Christopher R. Webster[1], Michael A. Jenkins[2], Shibu Jose[3]
and Linda M. Nagel*

[1] Ecosystem Science Center, School of Forest Resources and Environmental Science,
Michigan Technological University, Houghton, USA
[2] Department of Forestry and Natural Resources,
Purdue University, West Lafayette, IN USA
[3] School of Forest Resources and Conservation, 351 Newins-Ziegler Hall,
University of Florida, Gainesville, USA

ABSTRACT

A growing body of evidence suggests that exotic herb layers in forest understories are fundamentally altering ecological processes and successional dynamics. The establishment of mono-dominant patches of exotic plants is often influenced by complex interactions between propagule pressure, disturbance, climate, and land-use legacies. We use a combination of field studies and an extensive literature review to explore invasion ecology and consequences of invasion for the perpetuation of native forests along a gradient extending from southern Florida to the northern Lake States. Based on the commonalities between these invasions, we propose a general framework for integrating invasive species detection and control into forest management activities.

INTRODUCTION

Although dominated by species of seasonal importance and limited stature, herbaceous understories have the potential to strongly influence the structure and composition of forest

---

[1] Address: 1400 Townsend Drive, Houghton, MI 49931
[2] Address: 715 West State Street, West Lafayette IN 47907.
[3] Address: PO Box 110410, University of Florida, Gainesville, FL 32611-0410.

communities. The influence of the understory is most pronounced in its role as a selective filter that determines the spatial pattern, density, and species composition of forest regeneration (George and Bazzaz 2003). Forest seedling banks are a critical element for post-disturbance forest development, and a lack of advanced regeneration may slow stand development after canopy disturbance (Nakashizuka 1987). Advanced regeneration and overall population growth are dependent upon rates of transition from seed production to seedling establishment and growth, and from seedling establishment to sapling persistence and growth (Clark et al. 1999). Seedling establishment is frequently one of the strongest filters on recruitment (Clark et al. 1998; Gordon and Rice 2000), and herbaceous vegetation is often a major constituent of this filter (George and Bazzaz 2003).

Following disturbance, intense competition for above and belowground resources often occurs between the seedlings and sprouts of regenerating overstory species and resident herbaceous species (Jose et al. 2006; Gilliam 2007). The outcome of this competition has a lasting influence on the structure and composition of the developing stand. Much of the research that has examined competition between herbaceous vegetation and woody species regeneration has focused on the interaction between ferns and trees (Horsley 1977; Horsley 1993; George and Bazzaz 1999; de la Cretaz and Kelty 2002; Slocum et al. 2004). Research by Horsley (1993) found that competition for light is the primary mechanism of tree seedling inhibition employed by *Dennstaedtia punctilobula* (hay-scented fern). However, Lyon and Sharpe (2003) found that oak seedlings grown in competition with *D. punctilobula* had diminished content of Nitrogen (N), Potassium (K), and Phosphorus (P) than seedlings grown without fern competition, suggesting that belowground competition for nutrients may be occurring as well.

Other studies have shown that herbivory by *Odocoileus virginianus* (white-tailed deer) may act in conjunction with fern cover to suppress woody species regeneration (Horsley and Marquis 1983; de la Cretaz and Kelty 2002). In the Appalachian Mountains, other species that have been observed inhibiting hardwood regeneration include ferns such as *Thelypteris noveboracensis* (New York fern), *Pteridium aquilinum* (bracken fern), *Dryopteris carthusiana* (spinulose wood fern) and *Osmunda claytoniana* (interrupted fern) (Horsley and Marquis 1983; Maguire and Forman 1983; George and Bazzaz 1999), and grasses such as *Brachyelytrum erectum* (short husk grass) and *Danthonia compressa* (wild oat grass) (Horsley 1977; Bowersox and McCormick 1987).

While the potential of native herbaceous species to serve as ecological filters in forest understories has been clearly demonstrated, exotic herbaceous species have even greater potential to serve as filters through the inhibition of woody species regeneration. While invasive exotic plants may occur in relatively low densities in their native ranges, they often are able to achieve extremely high population densities in their introduced range. While only a small proportion of introduced species exhibit a capacity for extremely rapid population growth and long-term persistence, those that do have great ability to influence ecosystem structure and function (Davis 2003). For example, in New Zealand, Standish et al. (2001) observed that native forest seedling density and species richness decreased exponentially with increasing biomass of *Tradescantia fluminensis* (a perennial herb native to South America). In Hawaii, *Schizachyrium condensatum* and other introduced perennial grasses have had a strong negative effect on the recruitment of native woody species in woodlands. However, *S. condensatum* has been shown to slow the invasion of *Myrica fava* (a fast growing N-fixing tree introduced from the Canary Islands) into Hawaiian woodlands (D'Antonio and Mack

2001). Similarly, the invasive grass *Melinis minutiflora* reduced the regeneration of native woody species by more than half in savannas and forests of the Cerrado Region of Brazil (Hoffmann and Haridasan 2008).

As a general rule, the invasion of forest herbs into forest understories follows the inhibition model described by Connell and Slatyer (1977). In this model, an invasive species establishes in a community by securing growing space and then, in turn, inhibits the invasion of new species and suppresses the growth of those species already present. The release of exotic plants from coevolved herbivores has been offered as an important mechanism that allows exotics to out-compete native species and dominate communities (Blossey and Nötzold 1995; Williamson 1996; Keane and Crawley 2002). While research has documented the superior competitive ability of exotic plants (Dillenburg et al. 1993; Vilá and Weiner 2004), recent research has found that some exotic plants may also employ direct inhibition of competitors in addition to aggressive competition for resources. Numerous herbaceous species have been shown to chemically suppress fungal mutualists of native competitors (Stinson et al. 2006; Callaway et al. 2008; Wolfe at al. 2008), exude phytotoxins that directly inhibit the root growth of competitors (Bais et al. 2003; Callaway and Ridenour 2004), and alter microbial community function and structure in the soil (Kourtev et al. 2002; Ehrenfeld 2003; Wolfe and Klironomos 2005). Further, exotic grasses have been shown to maintain dominance by shifting native fire regimes towards more severe and frequent fires (D'Antonio and Vitousek 1992; Jose et al. 2002; Hoffmann et al. 2004) and exploiting interactions between drought and native herbivores (Webster et al. 2008). In this chapter, we present a review of three invasive herbaceous species that employ different mechanisms to suppress the regeneration of woody plants in forest stands. Our first example, *Alliaria petiolata* [Bieb] Cavara and Grande (garlic mustard) is a Eurasian herb that displaces native competitors through several mechanisms, including inhibition of ectomycorrhizal fungi (Callaway et al. 2008; Wolfe et al. 2008), subsequently affecting woody regeneration. Our second example, *Microstegium vimineum* (Trin.) A. Camus (Japanese stilt grass, Japanese grass, Nepal grass), is a shade-tolerant grass that persists in low light conditions (Claridge and Franklin 2002), alters soil microbial communities (Kourtev et al. 2002), and maintains dominance through its lack of palatability to white-tailed deer (*Odocoileus virginianus*) (Webster et al. 2008). Our final example, *Imperata cylindrica* (L.) Beauv. (Cogongrass), is a perennial grass that displaces its native competitor *Aristida stricta* (wiregrass) and creates a more-severe burning regime that prevents the regeneration of woody seedlings (Lippincott 2000). Based on the commonalities among these examples, we propose a general framework for integrating invasive species detection and control into forest management activities.

CASE STUDIES

Alliaria Petiolata

Alliaria petiolata (Figure 1) is one of the most widespread and problematic exotic invasive herbaceous species found in the understory of forests in eastern and central North America (Figure 2; Blossey et al. 2002).

Figure 1. *Alliaria petiolata* in flower (a.; Photo credit:Christopher R. Webster) and a heavily infested understory at Indiana Dunes State Park, IN (b.; Photo credit: Lindsey Shartell).

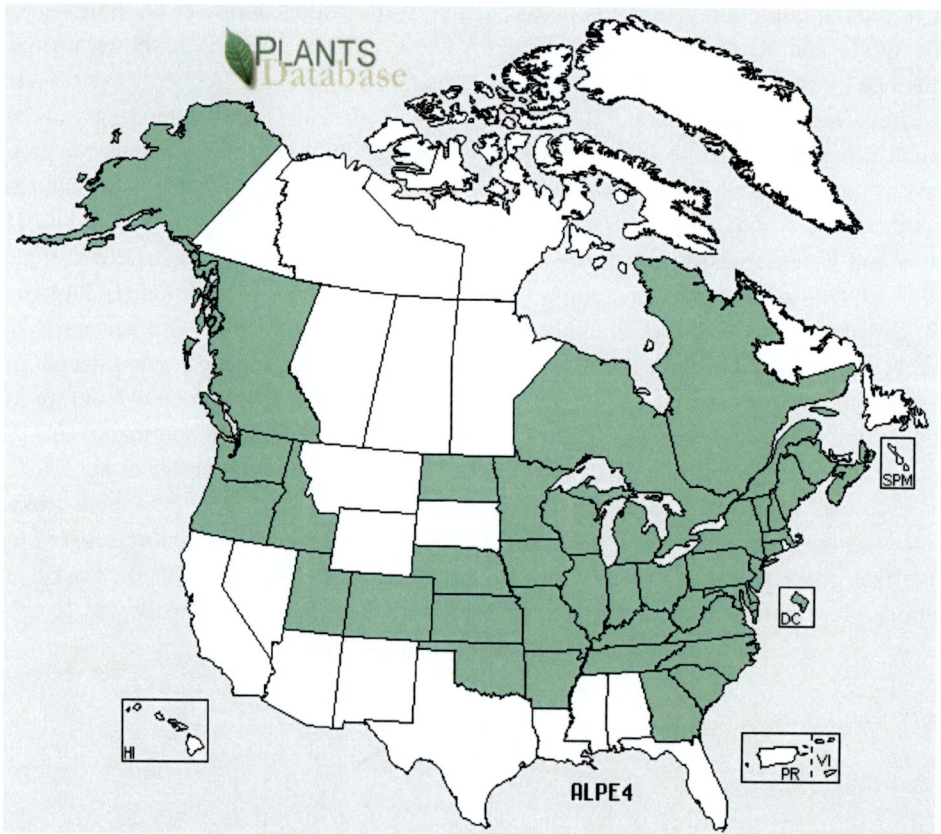

Figure 2. Current distribution of *Alliaria petiolata* (US Department of Agriculture Plants Database; http://plants.usda.gov/java/profile?symbol=ALPE4). States shaded in green indicate presence of the species.

The species is native to northern Europe, ranging from England to Sweden to the western part of the former USSR, and south to Italy (Nuzzo 1993). It has spread to North Africa, India, Sri Lanka, Canada, and the United States. It was first documented in the US in 1868 at Long Island, New York and in 1879 in Toronto, Canada (Nuzzo 1993; Anderson et al. 1996; Blossey et al. 2001). It is thought to have been introduced as a medicinal and edible plant. Where A. petiolata establishes, it can effectively eliminate other native vegetation, suppressing hardwood tree regeneration and impacting ecosystem function (Baskin and Baskin 1992; Nuzzo 1993). A. petiolata is easily identifiable by its heart-shaped, coarsely toothed leaves, white flowers (Figure 1a.), and seeds in long, slender siliques. It is an obligate biennial herb in the mustard family (Brassicaceae), forming rosettes that over-winter and emerge as flowering adults the following spring. Second-year plants grow rapidly in the spring while most native plants are still dormant (Anderson et al. 1996), with adult plants achieving heights well over one meter. Flowers are typically pollinated by insects, but the plant is also capable of self-pollination (Cruden et al. 1996). A single robust plant can have as many as 7,900 seeds with an average of 350 seeds per plant (Nuzzo 1993). Seeds of A. petiolata plants can mature and be viable when the plant is separated from the roots (Mortensen 2001, Frey 2005), and seedbanking for at least seven years has been documented (Baskin and Baskin 1992; Drayton and Primack 1999). Seeds can be expelled from the siliques up to two meters from the parent plant (Nuzzo 1999), though most seeds germinate within one meter of the parent plant, creating dense patches of emergent A. petiolata that are successful at crowding out native vegetation (Drayton and Primack 1999). An average rate of spread of 5.4 m per year was calculated in a "high quality, relatively undisturbed forest" with spread occurring faster into suitable microsites and slower into less suitable sites (Nuzzo 1999). Vectors and pathways for seed dispersal into forest understories include humans, animals, waterways, irrigation systems, roadways, trails, and lakeshores (Cavers et al. 1979).

A. petiolata commonly inhabits mesic forested areas dominated by mature hardwoods (Meekins and McCarthy 1999; Myers and Anderson 2003), and is more likely to invade species-rich rather than species-poor sites (Blossey et al. 2002). Though less competitive on other sites, A. petiolata can survive in wetland habitats, moist woods, and swamp forests (Voss 1985) as well as well-drained sunny sites (Meekins and McCarthy 2002). Increased alkalinity and nutrient content appear beneficial to growth of A. petiolata with its distribution associated with calcareous soils (Welk et al. 2002) and a noticeable absence from sites with acidic soils (Nuzzo 1991). This is consistent with its native range where A. petiolata grows best on base-rich soils (Cavers et al. 1979).

A. petiolata is also commonly found in disturbed areas, along roads, railroads, rivers, and urban areas (Nuzzo 1993). Invasion is often facilitated by both human and natural disturbances (Anderson et al. 1996; Welk et al. 2002), with repeated disturbances increasing the rate of spread (Nuzzo 1999). Though most often associated with disturbance (Byers and Quinn 1998), A. petiolata can also invade relatively undisturbed forests as well (Anderson et al. 1996). It often forms dense, distinct patches where it successfully invades (Winterer et al. 2005) resulting in overall higher productivity (including seed production) compared to areas where control efforts have reduced the density of A. petiolata (Rebek and O'Neil 2006).

A. petiolata is a successful invader due to a complex of additional growth characteristics and ecosystem effects within its introduced range. These characteristics, in turn, enable A. petiolata to suppress the regeneration of woody species regeneration. A study of leaf photosynthetic performance and leaf characteristics indicate high plasticity in A. petiolata under varying levels of shade, potentially increasing successful invasion and persistence in eastern forest understories (Myers et al. 2005). The species exhibits maximum photosynthetic rates early in the growing season before native spring ephemerals and summer forbs commence active growth (Myers and Anderson 2003). It is also thought to be a stronger competitor in its introduced range as compared to its home range due to biochemical effects. A non-mycorrhizal herbaceous plant, A. petiolata produces an allelochemical, benzyl isothiocyanate, that has been shown to inhibit the growth of ectomycorrhizal fungi (Wolfe et al. 2008) and affects associations of mycorrhizae with native tree seedlings, affecting their growth (Stinson et al. 2006). Further, suppression of mycorrhizal fungi corresponds with inhibition of plants that rely on these fungi in North America with correspondingly weak effects on similar European species (Callaway et al. 2008). When compared to its home range in Europe, A. petiolata in northeastern forests of North America exhibit lower herbivore damage, suggesting enemy release, though this was not accompanied by lower concentrations of sinigrin, a reputed defense compound (Lewis et al. 2006). A. petiolata also has allelopathic effects that directly affect other plant species (Prati and Bossdorf 2004). Additionally, cyanide is produced in the roots and aboveground tissues at levels that are toxic to many vertebrates (Cipollini and Gruner 2007). This may result in reduced herbivory of A. petiolata, offering it a further competitive advantage over frequently browsed woody seedlings.

Once established, A. petiolata has proven difficult to manage. Treatment methods have included hand-pulling, cutting, herbicide application, burning, and shading. Biological control agents under consideration include four weevil species from the genus Ceutorhynchus (Coleoptera: Curculionidae), which attack rosettes, stems, and seeds (Davis et al. 2006). Treatments are commonly applied at small scales, though little experimental data has been collected measuring effectiveness at controlling the invasive, or on impacts treatments have on native plant populations. In one study, spring treatment of glyphosate or mid-intensity fire proved effective in controlling adult and seedling density with fall treatments showing limited effectiveness (Nuzzo 1991). Drayton and Primack (1999) pulled flowering plants for four years, resulting in a decrease in population growth in 75% of the plots measured, though as in the former study, effects on native vegetation were not reported. A study in Upper Michigan where A. petiolata is not yet ubiquitous, showed a significant decrease in adult plants following herbicide, fire, and combination pull-herbicide and pull-scorch treatments, though these treatments appeared to increase seedling abundance one-year post-treatment as the plots revegetated (Nagel, unpublished data). The herbicide treatments also had an overall negative impact on species diversity one year post-treatment (Nagel, unpublished data). Application of glyphosate during early-season cold temperature shows promise. Frey et al. (2007) demonstrated 87 to 94% mortality of A. petiolata rosettes while nontarget native plants remained uninjured, and further, showed an increase in spring densities as compared to non-treated plots. General management recommendations involve exhausting the seedbank by preventing seed production through cutting, herbicide application, or prescribed fire. However, given its high fecundity and lengthy time seeds remain viable in the seedbank, it is unclear how many growing seasons of control are necessary, or if eradication is possible.

Microstegium Vimineum

Capable of forming dense monocultures, *Microstegium vimineum* (Figure 3) is a frequent invader of disturbed understory habitats along roads and streams and within floodplain and mesic forests of the eastern United States (Figure 4; Redman 1995; Cole and Weltzin 2004). First recorded in Knoxville, TN in 1919, *M. vimineum* appears to have been inadvertently introduced to North America from Asia as a result of its use as a packing material (Fairbrothers and Gray 1972; Barden 1987). This shade-tolerant, annual, C_4 grass forms a persistent seed bank, which in conjunction with its highly plastic morphological response to local microsite conditions enables it to hold invaded sites nearly indefinitely (Claridge and Franklin 2002; Gibson et al. 2002). In addition to replacing and excluding native plants (Barden 1987; Cole and Weltzin 2004; Oswalt et al. 2007), recent work also suggests that *M. vimineum* may significantly alter soil microbial communities (Kourtev et al. 2002).

Figure 3. Dense understory layer of Microstegium vimineum in Great Smoky Mountains National Park, TN, USA (Photo credit: Christopher R. Webster).

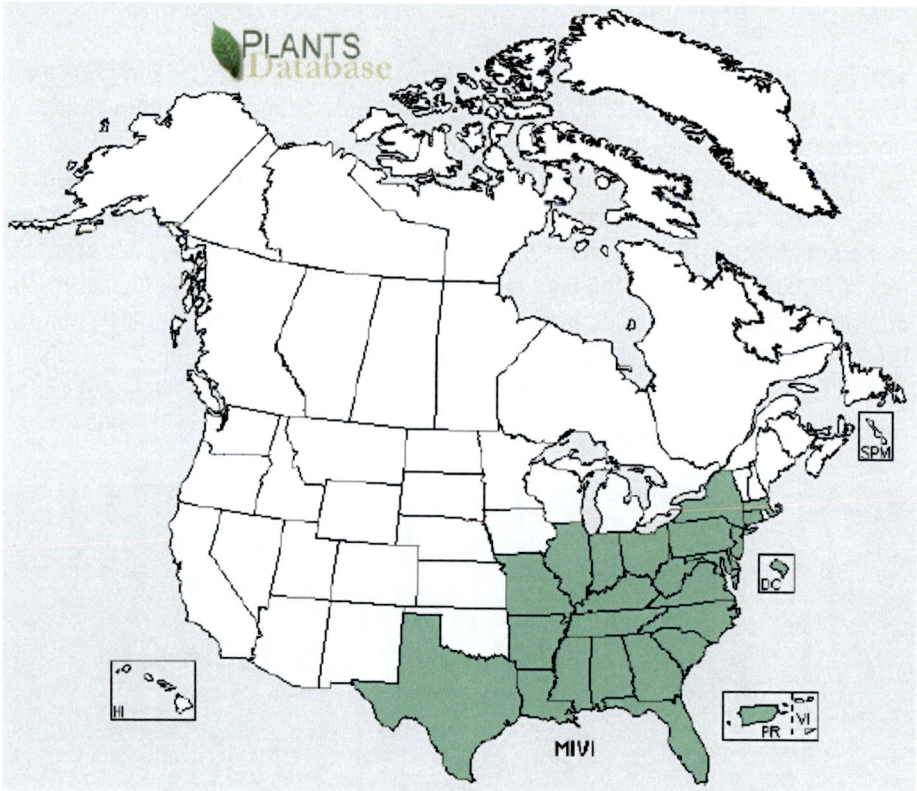

Figure 4. Current distribution of *Microstegium vimineum* (US Department of Agriculture Plants Database; http://plants.usda.gov/java/profile?symbol=MIVI). States shaded in green indicate presence of the species.

M. vimineum invades rapidly following disturbance of the litter layer in forest understories (Glasgow and Matlack 2007; Oswalt and Oswalt 2007). For example, Oswalt and Oswalt (2007) found that winter litter disturbance resulted in a rate of spread 4.5 times greater than on plots without disturbance. Additionally, litter disturbance resulted in a dramatic increase in standing biomass and spread even in the absence of overstory disturbance (Oswalt and Oswalt 2007). Canopy disturbance may further enhance invasion (Gibson et al. 2002), especially if it occurs in concert with understory disturbance (Glasgow and Matlack 2007). Given its propensity for disturbed microsite conditions, logging/skid and even animal (e.g., white-tailed deer) trails may serve as conduits for the invasion of *M. vimineum* (Oswalt and Oswalt 2007). This species also opportunistically invades from edge environments following even minor canopy disturbances, such as the death of a single tree (Cole and Weltzin 2004). The rapid and at times dramatic response of *M. vimineum* to disturbances related to forest management practices, such as harvesting and prescribed burning, has prompted a great deal of concern amongst natural resource mangers (Glasgow and Matlack 2007; Oswalt and Oswalt 2007).

As the coverage of *M. vimineum* increases the density and diversity of woody species in the ground-layer declines significantly (Oswalt et al. 2007). Once established, few if any native species appear to be able to perforate *M. vimineum* patches. Consequently, this species represents a strong ecological filter in the forest understory and can inhibit the perpetuation of tree species following canopy disturbance (see Figure 4 in Oswalt et al. 2007). Nevertheless, it is unclear how stable these patches are over the long-term. While growth and reproduction in this species are strongly influenced by moisture availability (e.g., drought; Gibson et al. 2002; Yurkonis and Meiners 2006; Webster et al. 2008), annual population declines and seed production failures as a result of severe drought appear to be buffered by the species persistent seed bank (Gibson et al. 2002). For example, after a severe drought in southern Illinois reduced flowering and seed production, Gibson et al. (2002) observed rapid reestablishment of *M. vimineum* populations the next year. The resilience of these populations was attributed to the species large, persistent seed bank at the site and seed dispersal from plants growing in more favorable microsites (Gibson et al. 2002). During drought years, flowering may be restricted to individuals growing in high light environments (Gibson et al. 2002).

In a recent study, Webster et al. (2008) examined the long-term outcome of interactions between drought and deer herbivory on the impermeability of *M. vimineum* patches to native tree species and forest herbs. Results suggest a close coupling of *M. vimineum* cover and drought severity (Webster et al. 2008). Surprisingly, in the absence of deer herbivory, native tree species were able to capitalize on transient declines in *M. vimineum* abundance. Following 10 years of deer exclusion, punctuated by a drought induced nadir in *M. vimineum* cover, native tree seedlings were recruiting into the sapling layer from areas formerly dominated by *M. vimineum*; whereas, no seedlings were able to emerge from the *M. vimineum* dominated field-layer on control plots (Webster et al. 2008). This work suggests that the impermeability of *M. vimineum* patches may be reinforced by selective browsing of native species emerging from an otherwise unpalatable field-layer (Webster et al. 2008). Emerging woody plants may induce a negative feedback on *M. vimineum* if they further reduce light levels near the forest floor (Cole and Weltzin 2005).

Once established, *M. vimineum* is difficult to eradicate, owing to its ease of reinvasion and persistent seed bank. Hand removal or mowing prior to seed set may provide control of small infestations, but are impractical for the control of large invasions (see Figure 3). The use of selective pre and/or post emergent herbicides may provide a more cost effective alternative to manual control. Currently, only one herbicide is specifically labeled for use against *M. vimineum*, but several formulations are being evaluated, with some showing significant promise (Judge et al. 2005). Nevertheless, the choice of control strategy should consider the specificity of the herbicide, the stage of invasion, and degree of degradation to the site. In other words, the approach used in a diverse understory recently invaded by *M. vimineum* should not further degrade the native plant community or seed bank. Conversely, in an area with a well-established *M. vimineum* mat, it may be necessary to neutralize the seed bank through the combined use and possible repeated application of pre- and post-emergent herbicides, after which native herbs may need to be actively reintroduced to the site.

Imperata Cylindrica

Imperata cylindrica (Figure 5) originated in southeast Asia and occurs throughout tropical and warmer regions of the world, from Japan to Southern China, throughout the Pacific islands, Australia, India, East Africa and the Southeastern United States (Holm et al. 1977).

Figure 5. Dense monotypic stand of *Imperata cylindrical* near Gainesville, FL (Photo credit: Shibu Jose).

In the southern United States, it has been shown to displace native understory plants, thereby threatening the ecological integrity of many invaded ecosystems (Collins et al. 2007). Imperata cylindrica was accidentally introduced to Alabama as a packing material in boxes from Japan in 1912 (Dickens 1974). It was then purposefully brought to Mississippi as a potential forage in 1921 (Dickens and Buchanan 1975). The unpalatability of I. cylindrica due to its high silica content prevented the use of it as a long-term forage (Dozier et al. 1998). Imperata cylindrica was also used to stabilize soil along roadways by state departments of transportation and spread throughout the southeastern United States through the use of rhizome infested soil in the construction and maintenance railroads (Jose et al. 2002). Currently, I. cylindrica is on the noxious weed list, which prohibits new plantings. It is also included in the Florida Department of Agriculture and Consumer Service's Noxious Weed List (Florida Statutes, Chapter 5B-57.007, 1993 revision) and the Florida Exotic Pest Council's 2003 invasive plant list. Imperata cylindrica is considered to be one of the ten most problematic and troublesome weeds in the world (MacDonald 2004).

Imperata cylindrica is reportedly established on over 500 million hectares of forested and agricultural land worldwide (Holm et al. 1977; Dozier et al. 1998; MacDonald 2004). In the U.S, I. cylindrica occurs in Alabama, Florida, Georgia, Mississippi, Louisiana, South Carolina, and Texas (Figure 6). It is an aggressive rhizomatous C_4 grass that can thrive with annual rainfall between 75 and 500 cm (Bryson and Carter 1993). It is a perennial grass with basal leaf blades that can be up to 1.5 m tall and 2 cm wide (Lippincott 1997). Leaf blades have a noticeably off-center whitish mid-vein and scabrous margins. The serrated margins of the leaves accumulate silicates, which deter herbivory (Dozier et al. 1998).

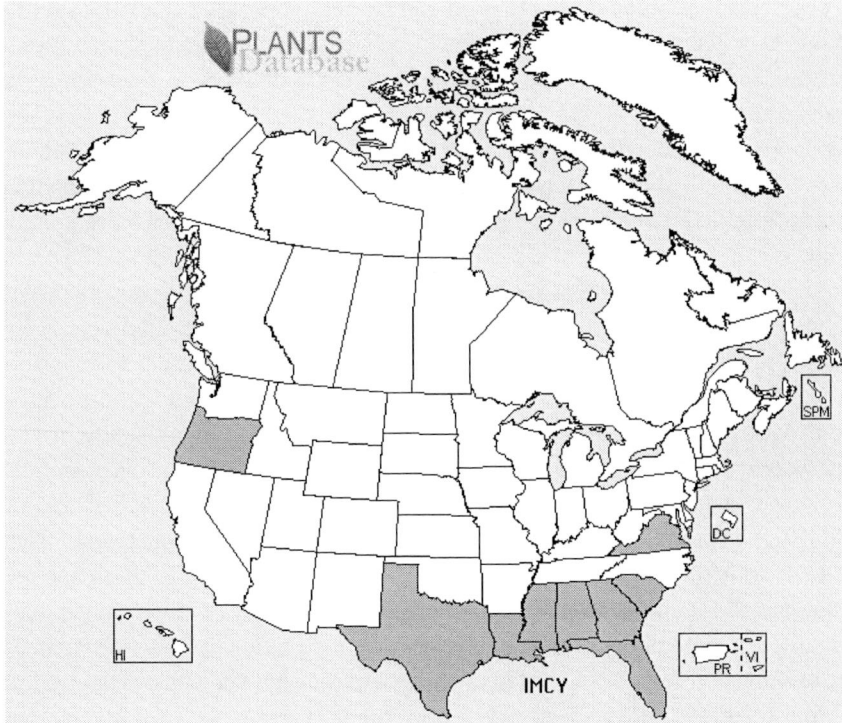

Figure 6. Current distribution of *Imperata cylindrica* (US Department of Agriculture Plants Database; http://plants.usda.gov/java/profile?symbol=IMCY). States shaded in green indicate presence of the species.

The rhizomes can comprise over 60% of the total plant biomass. The large belowground rhizome network leads to a low shoot-to-root/rhizome ratio and contributes to I. cylindrica's rapid regrowth after burning and cutting (Sajise 1976; Ramsey et al. 2003; MacDonald 2004). Rhizomes are very resistant to heat and breakage and may penetrate the soil 1.2 m deep, but generally occur in the top 0.15 m in heavy clay soils and 0.4 m in sandy soils (Holm et al. 1977). Rhizomes are white and tough with shortened internodes and possess several anatomical features, such as cataphylls (scale leaves) and sclerenchymous fibers, that help resist breakage and disruption when the rhizomes are trampled or disturbed.

A sexual reproduction occurs by rhizomes and sexual reproduction occurs by seeds. Sexual reproduction requires out crossing because I. cylindrica cannot self-pollinate and the rate of successful outcross is low (Shilling et al. 1997). It is a prolific seed producer with over 3000 seeds per plant. Seeds are capable of being dispersed long distances ranging from 15m to 100m. First flowering occurs within one year of germination and seeds germinate soon after ripening. No seed dormancy mechanisms have been observed, however seeds are highly germinable in natural populations. Seeds less than three months old have the highest viability with rapid decline in seed viability over time with a complete loss of viability after one year (Shilling et al. 1997).

I. cylindrica can invade a wide variety of habitats, soil types and climates. It has colonized deserts, sand dunes, grasslands, forests, river margins, and swamps. It is tolerant of a variety of soil conditions, but grows most favorably in acidic pH (pH 4.7) (Wilcut et al. 1988), low fertility and low organic matter soils. It thrives in highly disturbed areas such as roadsides and reclaimed mine areas (Bryson and Carter 1993). Dozier et al. (1998) indicated that seedling establishment is favored under conditions of limited competition, such as disturbed sites, and further suggested seedlings would be unlikely to establish in areas with greater than 75% sod cover.

The ecological impacts of I. cylidrica invasion have received considerable attention in the recent past. Although I. cylidrica invasion has been associated with disturbed sites, infestations in established plantations and natural stands are becoming more common in the U.S. Southeast. In a recent study Collins et al. (2007) examined Elton's hypothesis with respect to I. cylidrica invasion. Charles Elton hypothesized that invasion resistance and compositional stability increase with diversity (Elton 1958). The biodiversity-invasibility hypothesis postulates that species-rich communities are less vulnerable to invasion because vacant niches are less common and the intensity of interspecific competition is more severe. Field studies were conducted at two sites, a logged site and an unlogged site in Florida. The logged site was under 17-year-old loblolly pine (Pinus taeda) prior to clear-cutting and the unlogged site was a natural longleaf pine (Pinus palustris) forest which is believed to have the highest floristic diversity outside the tropics, owing to its rich understory. Their results indicated that both the logged site and unlogged site showed no significant relationship between the rate of I. cylindrica spread and native plant species richness, functional richness, and cover of the invaded community. Increased species or functional richness may increase the use of resources; however, the extensive rhizome/root network possessed by I. cylindrica and its ability to thrive under shade may allow for its persistence in a diverse community. Ramsey et al. (2003) showed that I. cylindrica has a light compensation point of 32 to 35 umol m^{-2} s^{-1}, indicating the ability of the species to survive as an understory species.

I. cylindrica invasion can change the structure and function of invaded ecosystems. It exerts intense competition for light, water and nutrients with native understory and overstory species (Lippincott 1997, Daneshgar and Jose, 2008). It forms dense, aggressive and persistent stands that displace native vegetation and suppresses the regeneration of native species. Preliminary data (Jose, unpublished) showed that I. cylindrica could significantly reduce native understory species cover in longleaf pine forests (Figure 7). Yager (2007), in her studies of the longleaf pine-blue stem and longleaf pine-shrub communities in Mississippi, observed lower species diversity and abundance of herbaceous vegetation in I. cylindrica infestations compared to uninfested adjacent areas. Lippincott (1997) also reported reduced herbaceous and woody species cover within I. cylindrica patches compared to adjacent uninfested sandhill longleaf pine communities. The fact that I. cylindrica is able to survive and persist through several survival strategies including an extensive rhizome network, adaptation to poor soils, drought tolerance, prolific wind-disseminated seed production, fire adaptability, lack of pests and diseases, and high genetic plasticity (MacDonald 2004) provides it with a competitive advantage over the native understory species.

The dense above ground biomass of I. cylindrica not only prevents recruitment of other plants, but also changes properties of the litter and upper soil layers (Lippincott 1997, Collins and Jose, 2008). Collins and Jose (2008) reported significantly lower levels of NO_3^--N and K^+ in I. cylindrica patches compared to the surrounding native vegetation in a longleaf pine forest. Lower levels of these nutrients were attributed to the extreme ability of I. cylindrica to extract available resources from the area in which it was invading. These authors also observed that the soil of the I.

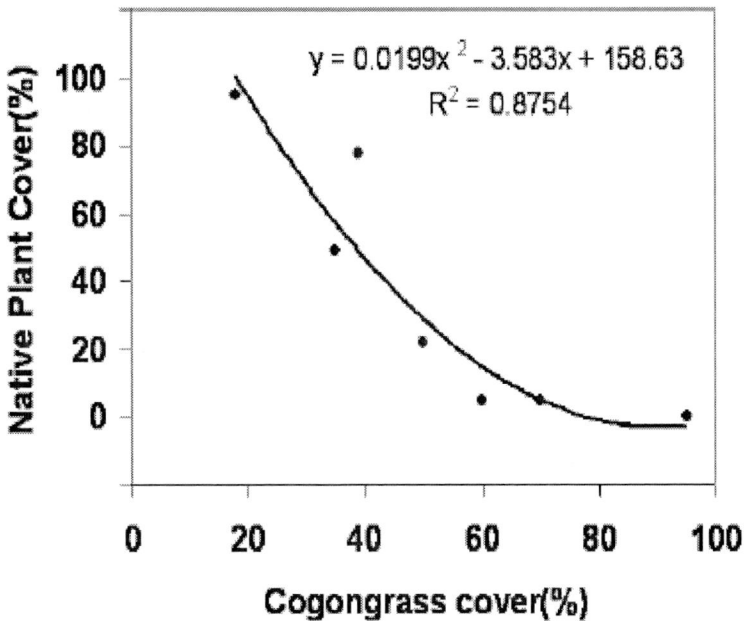

$$y = 0.0199x^2 - 3.583x + 158.63$$
$$R^2 = 0.8754$$

Figure 7. Native understory herbaceous cover as a function of Imperata cylindrica cover in a longleaf pine forest in Florida, USA.

I. cylindrica patch was more acidic than that of the surrounding native vegetation. Although they did not have direct evidence of any mechanisms responsible for lowering soil pH in I. cylindrica invaded patches, allelopathy or the preferential uptake of ammonium were suggested as plausible mechanisms. Based on research conducted by Koger and Bryson (2003), I. cylindrica appears to contain allelopathic substance(s) that contribute to its extreme invasive and competitive ability. Eussen and Soerjani (1975) and Eussen (1979) in a series of experiments showed that I. cylindrica suppressed the growth of tomato and cucumber and that the allelochemicals involved were more active at lower pH. The sensitivity of native understory species to allelochemicals from I. cylindrica, however, remains unknown.

Imperata cylindrica is a threat to ecosystems driven by fire. Dense monotypic stands of I. cylindrica disrupt fire regimes; fires become more intense and more frequent. Fire temperatures can reach 450°C at heights ranging from 0 to 1.5 m (Lippincott 2000). Increased fire intensity causes increased damage and death to seedlings and juvenile native species. Changes in fire regimes displace native plants and flora and decrease productivity of the stand as well. Although fire is a tool used in the management and restoration of the longleaf pine ecosystem, fire can accelerate the spread of I. cylindrica if control measures are not used. In a recent research study by Yager (2007), two longleaf pine communities (pine-bluestem and pine-shrub vegetation) were examined with respect to vegetative spread of I. cylindrica with and without burning. Mean vegetative encroachment of I. cylindrica was < 2 m/yr for both communities; however, vegetative spread was more than double in burned plots compared to unburned plots.

Past research has shown that alterations to the native fire regime caused by invasion of I. cylindrica infestation inhibits seedling recruitment of overstory tree species. Lippincot (1997) compared fire related mortality of longleaf pine seedlings and saplings in areas infested with

I. cylindrica and with native understory in longleaf pine forests and observed that mortality was significantly higher in the presence of I. cylindrica. Lippincot also observed that I. cylindrica burned hotter than the native understory species, which caused the significantly higher mortality of overstory trees, in addition to seedlings and saplings. As in Lippincott's (1997) study, Yager (2007) also observed lower abundance of longleaf seedlings in I. cylindrica infested areas.

Controlling I. cylindrica in the forest understory is challenging. The two most effective herbicides, imazapyr and glyphosate (Willard et al. 1997), can also have negative effects on other understory plants. Cultural treatments such as disking or burning do not effectively control I. cylindrica (Gaffney, 1996; Willard, 1998); however, they may be used to reduce seed production. Successful management of this grass requires killing of all rhizomes (Dozier et al., 1998). Application of herbicides in the fall before the first frost, when movement of photosynthate to the rhizomes takes place at a faster rate, has been shown to give the greatest control. Further, multiple applications are often needed to inhibit regrowth from the extensive rhizome system. Although continued disking can be beneficial it may not be practical in natural stands or plantations. An integrated management approach using all available methods such as burning, disking, mowing, herbicide applications, and revegetating the area (with natives capable of persisting with or inhibiting the reestablishment of I. cylindrica) is recommended as the key to achieving I. cylindrica control (Jose et al. 2002).

A FRAMEWORK FOR CONTROL

As illustrated by our review, several invasive exotic understory herbs have the capacity to become important ecological filers in forested ecosystems, altering both herbaceous-layer and future overstory communities. Of particular concern are species capable of disrupting historic developmental trajectories and fundamentally altering the potential of native forests for the sustainable production of fiber and other ecosystem services (i.e., "transformers" *sensu* Richardson et al. [2000]). Once these species have a secure foothold they are difficult, if not impossible, to eradicate. Therefore, the key attribute of any invasive species control program should be prevention. Prevention requires active monitoring for known invasives and risk assessment regarding activities that may facilitate invasion and sites that by their nature, spatial juxtaposition, or disturbance history are prone to invasion. Resources for monitoring will always be limiting, but waiting for these species to take hold before taking action is a much more costly endeavor. Additionally, cooperation between landowners, both private and public, will be required to address local and regional source populations that contribute to high rates of invasion and reinvasion.

Given the potential consequences of invasion for the sustainable management and ecological stability of native forest ecosystems, the most effective technique available should be deployed as rapidly as possible following detection. This approach has been termed "Early Detection, Rapid Response" and is based on recommendations by the Federal Interagency Committee for the Management of Noxious and Exotic Weeds (Webster et al. 2006). Conceptually, this approach builds on work by Moody and Mack (1988) who demonstrated in a modeling environment that the most effective way to arrest the spread of an invasive species is to first focus on satellite populations (nascent foci) distal from the primary invasion front.

In areas with well established populations of aggressive exotic herbs, this approach can be readily applied following control activities to help prevent reinvasion. Control activities in these areas should be incorporated into timber stand improvement and site preparation treatments. Ignoring established populations won't make them go away, and a light infestation can rapidly transition into an impervious field layer following site disturbance resulting from harvesting or natural disturbance (Oswalt et al. 2007). In situations where broadcast control is not feasible, spot control may be sufficient to facilitate successful regeneration of native tree species. Initial control activities in heavily infested sites may focus specifically on areas where regeneration is desired such as canopy gaps and harvest openings. As shown for *M. vimineum*, herbivore exclusion may also be necessary so that trees and herbs emerging from an otherwise unpalatable field layer are not lost to browsing.

In heavily infested areas, it should not be presumed that effective control or even eradication of the invader will result in plant community recovery. Depending on time since invasion and the treatments used to control the invader, there may be few native propagules in place to reoccupy the site. Denuded understories are highly susceptible for reinvasion and the site disturbance associated with treatment in some cases may actually accelerate invasion. Consequently, control measure may need to be coupled with equally aggressive measures aimed at enhancing or re-introducing native vegetation. This may be especially important in areas where eradication is impossible and the goal is to maintain a viable component of native species.

REFERENCES

Anderson, R.C., Dhillion, S.S. and Kelley, T.M. (1996) Aspects of the ecology of an invasive plant, garlic mustard (*Alliaria petiolata*) in central Illinois. *Restoration Ecology*, 4, 181-191

Bais, H.P., Vepachedu, R., Gilroy, S., Callaway, R.M. and Vivanco, J.M. (2003) Allelopathy and exotic plant invasions: from molecules and genes to species interactions. *Science*, 301, 1377-1380

Barden, L.S. (1987) Invasion of *Microstegium vimineum* (Poaceae), an exotic, annual, shade-tolerant, C_4 grass, into a North Carolina floodplain. *American Midland Naturalist*, 118, 40-45

Baskin, J. M. and Baskin, C.C. (1992) Seed germination biology of the weedy biennial *Alliaria petiolata*. *Natural Areas Journal*, 12, 191-197

Blossey, B. and Nötzold, R. (1995) Evolution of increased competitive ability in invasive nonindigenous plants: a hypothesis. *Journal of Ecology*, 83, 887-889

Blossey, B., Nuzzo, V., Hinz, H. and Gerber, E. (2001) Developing biological control of *Alliaria petiolata* (M. Bieb.) Cavara and Grande (garlic mustard). *Natural Areas Journal*, 21, 357-367

Blossey, B., Nuzzo, V., Hinz, H. and Gerber, E. (2002) Garlic Mustard *In:* Van Driesche, R., B. Blossey, M. Hoddle, S. Lyon, and R. Reardon (eds.). *Biological Control of Invasive Plants in the Eastern United States*. pp 365-372. U.S. Department of Agriculture, Forest Service, FHTET-2002-04

Bowersox, T.W.and McCormick, L.H. (1987) Herbaceous communities reduce the juvenile growth of northern red oak, white ash, yellow poplar, but not white pine. pp. 39-42, *in* Hay, R.L., Woods, F.W. and DeSelm, H. (eds.). *Proceedings of the Sixth Central Hardwood Forest Conference*, February 24-26, 1987, Knoxville, TN

Bryson, C.T. and Carter, R. (1993) Cogongrass, *Imperata cylindrica*, in the United States. *Weed Technology*, 7, 1005-1009

Byers, D.L. and Quinn, J.A. (1998) Demographic variation in *Alliaria petiolata* (Brassicaceae) in four contrasting habitats. *Journal of the Torrey Botanical Society*, 125, 138-149

Callaway, R.M., Cipollini, D., Barto, K. Thelen, G.C., Hallett, S.G., Prati, D., Stinson, K. and Klironomos, J. (2008) Novel weapons: invasive plants suppress fungal mutualists in America but not in its native Europe. *Ecology*, 89, 1043-1055

Callaway, R.M. and Ridenour, W.M. (2004) Novel weapons: invasive success and the evolution of increased competitive ability. *Frontiers in Ecology and the Environment*, 2, 436-443

Cavers, P.B., Heagy, M.I. and Kokron, R.F. (1979) The biology of Canadian weeds. 35. *Alliaria petiolata* (M. Bieb.) Cavara and Grande. *Canadian Journal of Plant Science*, 59, 217-229

Cipollini, D. and Gruner, B. (2007) Cyanide in the chemical arsenal of garlic mustard, *Alliaria petiolata*. *Journal of Chemical Ecology*, 33, 85-94

Claridge, K. and Franklin, S.B. (2002) Compensation and plasticity in an invasive species. *Biological Invasions*, 4, 339-347

Clark, J.S., Beckage, B., Camill, P., Cleveland, B., Hillerislambers, J., Lichter, J. McLachlan, J., Mohan, J. and Wycoff, P. (1999) Interpreting recruitment limitations in forests. *American Journal of Botany*, 86, 1-16

Clark, J.S., Macklin, E. and Wood, L. (1998) Stages and spatial scales of recruitment limitations in southern Appalachian forests. *Ecological Monographs*, 68, 213-235

Cole, P.G. and Weltzin, J.F. (2004) Environmental correlates of the distribution and abundance of *Microstegium vimineum*, in east Tennessee. *Southeastern Naturalist*, 3, 545-562

Cole, P.G. and Weltzin, J.F. (2005) Light limitation creates patchy distribution of an invasive grass in eastern deciduous forests. *Biological Invasions*, 7, 477-488

Collins, A.R., Jose, S., Daneshgar, P. and Ramsey, C. (2007) Elton's hypothesis revisited: *Imperata cylindrica* invasion in the southeastern United States. *Biological Invasions*, 9, 433-443

Connell, J.H. and Slatyer, R.O. (1977) Mechanisms of succession in natural communities and their role in community stability and organization. *The American Naturalist*, 111, 1119-1144

Cruden, R.W., McClain, A.M. and Shrivastava, G.P. (1996) Pollination and breeding system of *Alliaria petiolata* (Brassicaceae). *Bulletin of the Torrey Botanical Club*, 123, 273-280

Daneshgar P. (2007) *Imperata cylindrica* invasion in juvenile *Pinus taeda* forests: implications of diversity and impacts on productivity and nitrogen dynamics. Ph.D. Dissertation, University of Florida, Gainesville, FL

D'Antonio, C.M. and Mack, M. (2001) Exotic grasses potentially slow invasion of an N-fixing tree into a Hawaiian woodland. *Biological Invasions*, 3, 69-73

D'Antonio, C.M. and Vitousek, P.M. (1992) Biological invasions by exotic grasses, the grass/fire cycle, and global change. *Annual Review of Ecology and Systematics*, 23, 63-87

Davis, M.A. (2003) Biotic globalization: does competition from introduced species threaten biodiversity? *BioScience*, 53, 481-489

Davis, A.S., Landis, D.A., Nuzzo, V., Blossey, B., Gerber, E. and Hinz, H.L. (2006) Demographic models inform selection of biocontrol agents for garlic mustard (*Alliaria petiolata*). *Ecological Applications*, 16, 2399-2410

de la Cretaz, A.L. and Kelty, M.J. (2002) Development of tree regeneration in fern-dominated forest understories after reduction of deer browsing. *Restoration Ecology*, 10, 416-426

Dickens, R. (1974) Cogongrass in Alabama after sixty years. *Weed Science*, 22, 177-179

Dickens, R. and Buchanan, G.A. (1975) Control of cogongrass with herbicides. *Weed Science*, 23, 194-197

Dillenburg, L.R., Whigham, D.F., Teramura, A.H and Forseth, I.N. (1993) Effects of below and above ground competition from the vines *Lonicera japonica* and *Parthenocissus quinquefolia* on the growth of the tree host *Liquidambar styraciflua*. *Oecologia*, 93, 48-54

Dozier, H., Gaffney, J.F., McDonald, S.K., Johnson, E.R.R.L. and Shilling, D.G. (1998) Cogongrass in the United States: History, Ecology, Impacts and Management. *Weed Technology*, 12, 737-743

Drayton, B. and Primack, R.B. (1999) Experimental extinction of garlic mustard (*Alliaria petiolata*) populations: implications for weed science and conservation biology. *Biological Invasions*, 1, 159-167

Ehrenfeld, J.G. (2003) Effects of Exotic Plant Invasions on Soil Nutrient Cycling Processes. *Ecosystems*, 6, 503-523

Elton, C.S. (1958) The Ecology of Invasions by Animals and Plants. T. Methuen and Co., London

Fairbrothers, D.E. and Gray, J.R. (1972) *Microstegium vimineum* (Trin.) A. Camus (Gramineae) in the United States. *Bulleting of the Torrey Botanical Club*, 99, 97-100

Frey, M. (2005) Spraying glyphosate at freezing temperatures and other techniques for controlling garlic mustard (Ohio). *Ecological Restoration*, 23, 280-281

Frey, M.N., Herms, C.P. and Cardina, J. (2007) Cold weather application of glyphosate for garlic mustard (*Alliaria petiolata*) control. *Weed Technology*, 21, 656-660

George, L.O. and Bazzaz, F.A. (1999) The fern understory as an ecological filter: growth and survival of canopy-tree seedlings. *Ecology*, 80, 846-856

George, L.O. and Bazzaz, F.A. (2003) The herbaceous layer as a filter determining spatial pattern in forest tree regeneration. pp. 105-159, *in* Gilliam, F.S. and Roberts, M.R. (eds.) *The Herbaceous Layer in Forests of Eastern North America*, Oxford University Press, New York, NY, US

Gibson, D.J., Spyreas, G. and Benedict, J. (2002) Life history of *Microstegium vimineum* (Poaceae), an invasive grass in southern Illinois. *Journal of the Torrey Botanical Society*, 129, 207-219

Gilliam, F.S. (2007) The ecological significance of the herbaceous layer in temperate forest ecosystems. *BioScience*, 57, 845-857

Glasgow, L.S. and Matlack, G.R. (2007) The effects of prescribed burning and canopy openness on establishment of two non-native plant species in a deciduous forest, southeast Ohio, USA. *Forest Ecology Management,* 238, 319-329

Gordon, D.R. and Rice, K.J. (2000) Competitive suppression of *Quercus douglasii* (Fagaceae) seedling emergence and growth. *American Journal of Botany*, 87, 986-994

Hoffmann, W.A. and Haridasan, M. (2008) The invasive grass, *Melinis minutiflora*, inhibits tree regeneration in a Neotropical savanna. *Austral Ecology*, 33, 29-36

Hoffmann, W.A., Lucatelli, V.M.P.C., Silva, F.J., Azeuedo, I.N.C., Marinho, M.S., Albuquerque, A.M.S., Lopes, A.O. and Moreira, S.P. (2004) Impacts of the invasive alien grass *Melinis minutiflora* at the savanna-forest ecotone in the Brazilian Cerrado. *Diversity and Distributions*, 10, 99-103

Holm, L.G., Plucknett, D.L., Pancho, J.V. and Herberger, J.P. (1977) The World's Worst Weeds: Distribution and Biology. University Press of Hawaii, Honolulu, Hawaii, USA

Horsley, S.B. (1977) Allelopathic inhibition of black cherry by fern, grass, goldenrod, and aster. *Canadian Journal of Forest Research*, 7, 205-216

Horsley, S.B. (1993) Mechanisms of interference between hay-scented fern and black cherry. *Canadian Journal of Forest Research*, 23, 2059-2069

Horsley, S.B. and Marquis, D.A. (1983) Interference of weeds and deer with Allegheny hardwood reproduction. *Canadian Journal of Forest Research*, 13, 61-69

Jose, S., Cox, J. Miller, D.L., Shilling, D.G. and Merritt, S. (2002) Alien plant invasions: the story of I. cylindrica in southeastern forests. *Journal of Forestry*, 100, 41-44

Jose, S., Williams, R.A. and Zamora, D.S. (2006) Belowground ecological interactions in mixed-species forest plantations. *Forest Ecology and Management*, 233, 231-239

Judge, C.A., Neal, J.C. and Derr, J.F. (2005) Preemergence and postemergence control of Japanese stiltgrass (M*icrostegium vimineum*). *Weed Technology,* 19, 183-189

Keane, R.M. and Crawley, M.J. (2002) Exotic plant invasions and the enemy release hypothesis. *Trends in Ecology and Evolution*, 17, 164-170

Koger, C.H. and Bryson, C.T. (2003) Effect of cogongrass (*Imperata cylindrica*) residues on bermudagrass (*Cynodon dactylon*) and Italian ryegrass (*Lolium multiflorum*). *Proceedings, Southern Weed Science Society*, 56, 341

Kourtev, P.S., Ehrenfeld, J.G. and Häggblom, M. (2002) Exotic plant species alter the microbial community structure and function in the soil. *Ecology,* 83, 3152-3166

Lewis, K.C., Bazzaz, F.A., Liao, Q. and Orians, C.M. (2006) Geographic patterns of herbivory and resource allocation to defense, growth, and reproduction in an invasive biennial, *Alliaria petiolata*. *Oecologia*, 148, 384-395

Lippincott, C.L. (1997) Ecological consequences on *Imperata Cylindrica* (cogongrass) invasion in Florida Sandhill. Ph.D. University of Florida, Gainesville

Lippincott, C.L. (2000) Effects of *Imperata cylindrica* (L.) Beauv. (Cogongrass) invasion of fire regimes in Florida Sandhill (USA). *Natural Areas Journal*, 20, 140-149

Lyon, J. and Sharpe, W.E. (2003) Impacts of hay-scented fern on nutrition of northern red oak seedlings. *Journal of Plant Nutrition*, 26, 487-502

MacDonald, G.E. (2004) Cogongrass (*Imperata cylindrica*) – Biology, Ecology and Management. *Critical Reviews in Plant Sciences*, 23, 367-380

Maguire, D.A. and Forman. R.T.T. (1983) Herb cover effects on tree seedling patterns in a mature hemlock-hardwood forest. *Ecology*, 64, 1367-1380

Meekins, J.F. and McCarthy, B.C. (1999) Competitive ability of *Alliaria petiolata* (garlic mustard, Brassicaceae), an invasive, nonindigenous forest herb. *International Journal of Plant Science*, 160, 743-752

Meekins, J.F. and McCarthy, B.C. (2002) Effect of population density on the demography of an invasive plant (*Alliaria petiolata*, Brassicaceae) population in a southeastern Ohio forest. *American Midland Naturalist*, 147, 256-278

Moody, M.E. and Mack, R.N. (1988) Controlling the spread of plant invasions: the importance of nascent foci. *Journal of Applied Ecology*, 25, 1009-1021

Mortensen, C.E. (2001) Garlic mustard in pocket guide, *In:* Weeds of the Northern Lake States. USDA Forest Service, Leech Lake (MN) Band of Ojibwe, US Department of the Interior-Bureau of Indian Affairs, Washington D.C.

Myers, C.V. and Anderson, R.C. (2003) Seasonal variation in photosynthetic rates influences success of an invasive plant, garlic mustard (*Alliaria petiolata*). *American Midland Naturalist*, 150, 231-245

Myers, C.V., Anderson, R.C. and Byers, D.L. (2005) Influence of shading on the growth and leaf photosynthesis of the invasive non-indigenous plant garlic mustard [*Alliaria petiolata* (M. Bieb) Cavara and Grande] grown under simulated late-winter to mid-spring conditions. *Journal of the Torrey Botanical Society*, 132, 1-10

Nakashizuka, T. (1987) Regeneration dynamics of beech forests in Japan. *Vegetatio*, 69, 169-175

Nuzzo, V.A. (1991) Experimental control of garlic mustard [*Alliaria petiolata* (Bieb.) Cavara and Grande] in northern Illinois using fire, herbicide, and cutting. *Natural Areas Journal*, 11, 158-167

Nuzzo, V.A. (1993) Distribution and spread of the invasive biennial *Alliaria petiolata* (garlic mustard) in North America *In:* McKnight, B.N. (ed.). *Biological Pollution: The Control and Impact of Invasive Exotic Species.* pp 137-146. Indiana Academy of Science, Indianapolis

Nuzzo, V.A. (1999) Invasion pattern of the herb garlic mustard (*Alliaria petiolata*) in high quality forests. *Biological Invasions*, 1, 169-179

Oswalt, C.M. and Oswalt, S.N. (2007) Winter litter disturbance facilitates the spread of the nonnative invasive grass *Microstegium vimineum* (Trin.) A. Camus. *Forest Ecology Management*, 249, 199-203

Oswalt, C.M., Oswalt, S.N. and Clatterbuck, W.K. (2007) Effects of *Microstegium vimineum* (Trin.) A. Camus on native woody species density and diversity in a productive mixed-hardwood forest in Tennessee. *Forest Ecology Management*, 242, 727-732

Prati, D. and Bossdorf, O. (2004) Allelopathic inhibition of germination by *Alliaria petiolata* (Brassicaceae). *American Journal of Botany*, 91, 285-288

Ramsey, C.L., Jose, S., Miller, D.L., Cox, J., Portier, K.M., Shilling, D.G. and Merritt, S. (2003) Cogongrass [*Imperata cylindrica* (L.) Beauv.] response to herbicide and disking on a cutover site in a mid-rotation pine plantation in Southern USA. *Forest Ecology and Management*, 179, 195-209

Rebek, K.A. and O'Neil, R.J. (2006) The effects of natural and manipulated density regimes on *Alliaria petiolata* survival, growth and reproduction. *Weed Research*, 46, 345-352

Redman, D.E. (1995) Distribution and habitat types for Nepal Microstegium [*Microstegium vimineum* (Trin.) A. Camus] in Maryland and the District of Columbia. *Castanea*, 60, 270-275

Richardson, D.M., Pyšek, P., Rejmánek, M., Barbour, M.G., Panetta, F.D. and West, C.J. (2000) Naturalization and invasion of alien plants: concepts and definitions. *Diversity and Distributions*, 6, 93-107

Royo, A.A. and Carson, W.P. (2006) On the formation of dense understory layers in forests worldwide: consequences and implications for forest dynamics, biodiversity, and succession. *Canadian Journal of Forest Research*, 36, 1345-1362

Sajise, P.E. (1976) Evaluation of cogon [*Imperata cylindrica* (L.) Beauv.] as a seral stage in Philippine vegetational succession. 1. The cogonal seral stage and plant succession. 2. Autoecological studies on cogon. Dissertation Abstracts International B. (1973) 3040-3041. *Weed Abstracts*, No. 1339.

Shilling, D.G., Beckwick, T.A., Gaffney, J.F., McDonald, S.K., Chase, C.A. and Johnson, E.R.R.L. (1997) Ecology, physiology, and management of Cogongrass (*Imperata cylindrica*). Florida Institute of Phosphate Research, Bartow, FL

Slocum, M.G., Aide, T.M., Zimmerman, J.K. and Navarro, L. (2004) Natural regeneration of subtropical montane forest clearing fern thickets in the Dominican Republic. *Journal of Tropical Ecology*, 20, 483-486

Standish, R.J., Robertson, A.W. and Williams, P.A. (2001) The impact of an invasive weed *Tradescantia fluminensis* on native forest regeneration. *Journal of Applied Ecology*, 38, 1253-1263

Stinson, K.A., Campbell, S.A., Powell, J.R., Wolfe, B.E., Callaway, R.M. Thelen, G.C., Hallett, S.G., Prati, D. and Klironomos, J.N. (2006) Invasive plants suppresses the growth of native tree seedlings by disrupting belowground mutualisms. *Public Library of Science Biology*, 4, 727-731

Vilá, M. and Weiner, J. (2004) Are invasive species better competitors than native plant species? – evidence from pair-wise experiments. *Oikos*, 105, 229-238

Voss, E.G. (1985) Michigan Flora. Part II Dicots. Cranbrook Institute of Science and University of Michigan Herbarium.

Webster, C.R., Rock, J.H., Froese, R.E. and Jenkins, M.A. 2008. Drought-herbivory interaction disrupts competitive displacement of native plants by *Microstegium vimineum*, 10 year results. *Oecologia*, 157, 497-508

Welk, E., Schubert, K. and Hoffmann, M.H. (2002) Present and potential distribution of invasive garlic mustard (*Alliaria petiolata*) in North America. *Diversity and Distributions*, 8, 219-233

Wilcut, J.W., Dute, R.R., Truelove, B. and Davis, D.E. (1988) Factors limiting the distributions of Cogongrass (*Imperata cylindrica*), and Torpedograss (*Panicum repens*). *Weed Science*, 36, 49-55

Willard, T.R., Gaffney, J.F. and Shilling, D.G. (1997) Influence of herbicides combinations and application technology on cogongrass (*Imperata cylindrica*) control. *Weed Technology*, 11, 76-80

Williamson, M. (1996) Biological Invasions, Chapman and Hall, London

Winterer, J., Walsh, M.C., Poddar, M., Brenna, J.W. and Primak, S.M. (2005) Spatial and temporal segregation of juvenile and mature garlic mustard plants (*Alliaria petiolata*) in a Central Pennsylvania woodland. *American Midland Naturalist*, 153, 209-216

Wolfe, B.E. and Klironomos, J.N. (2005) Breaking new ground: soil communities and exotic plant invasion. *BioScience*, 55, 477-487

Wolfe, B.E., Rodgers, V.L., Stinson, K.A. and Pringle, A. (2008) The invasive plant *Alliaria petiolata* (garlic mustard) inhibits ectomycorrhizal fungi in its introduced range. *Journal of Ecology*, 96, 777-483, doi: 10.1111/j.1365-2745.2008.01389.x

Yager, L.Y. (2007) Watching the grass grow: effects of habitat type, patch size, and land use on cogongrass (*Imperata cylindrica* (L.) Beauv.) spread on Camp Shelby Training Site Mississippi. Ph.D. Dissertation, Mississippi State University. 178 p

Yurkonis, K.A. and Meiners, S.J. (2006) Drought impacts and recovery are driven by local variation in species turnover. *Plant Ecology*, 184, 325-336

In: Forest Canopies: Forest Production, Ecosystem...   ISBN 978-1-60741-457-5
Editor: J. D. Creighton and P. J. Roney   © 2009 Nova Science Publishers, Inc.

Chapter 3

QUANTITATIVE ANALYSIS OF CANOPY PHOTOSYNTHESIS INFLUENCED BY LIGHT SIMULATION MODELS

Toru Sakai[1], Hiroyuki Muraoka[2], Tsuyoshi Akiyama[2], Michio Shibayama[3] and Yoshio Awaya[2]

1 Research Institute for Humanity and Nature, 457-4 Motoyama,
Kamigamo, Kita-ku, Kyoto, 603-8047 Japan
2 River Basin Research Center, Gifu University, 1-1 Yanagido,
Gifu, 501-1193, Japan
3 National Institute for Agro-Environmental Sciences,
3-1-3 Kannondai, Tsukuba, 305-8604, Japan

ABSTRACT

Light distribution pattern is a critical input for canopy photosynthesis models. We examined how and to what extent the different calculations of light distribution affect estimations of daily photosynthesis (A_{day}) of herbaceous vegetation, *Sasa senanensis*, in a deciduous forest understory. A_{day} was estimated using three models, the M-S$_1$ (the Monsi-Saeki model 1; diurnal incident light was defined by sine curve from sunrise to sunset and the vertical profile of incident light through the stand was calculated using the Beer's law), M-S$_2$ (a model combining the hemispherical photographic method to describe sunfleck/diffuse light above the stands and Beer's law to describe the vertical profile of incident light within the stands) and Y-plant (incident light was calculated for every single leaves by using a hemispherical photograph and geographical distribution of single leaf area and orientation). Although estimations of daily light absorption by the whole stands were relatively close among three models, A_{day} varied greatly in order of the M-S$_1$, M-S$_2$ and Y-plant models under clear sky conditions. The significant overestimation of A_{day} was attributed to the different manner of calculating (i) the diurnal pattern of incident light above the stands and (ii) the light distribution pattern within the stands. The use of

---
[1] Corresponding Author: Toru Sakai, Tel: +81-75-707-2457, Fax: +81-75-707-2513, E-mail: torus@chikyu.ac.jp

diurnal pattern of incident light estimated by a sine curve led to a 76% overestimation of A_{day} in the model separating sunflecks/diffuse light (i.e., the M-S_1 model vs the M-S_2 model). In addition, less calculation of the clumping effect such as leaf overlaps caused a 78% overestimation of A_{day}, and the error was larger with higher LAI (i.e., the M-S_2 model vs the Y-plant model). Consequently, A_{day} could be overestimated by an average of 213% as a result of ignoring dynamic responses of light. It is necessary to understand the possibilities and limitations of the proposed model and to determine which models make more efficient use of survey data. The application of such models often depends on the availability and quality of required data. The quantitative, comparative studies are important to gain a better understanding of physiological process. Our study showed the possible ranges how much A_{day} was affected by light distribution patterns, stand density and seasonal changes in the light environment.

Keywords: canopy photosynthesis; forest understory; leaf area index; light environment; model

INTRODUCTION

The measurement of CO_2 exchange between the vegetation and the atmosphere is becoming increasingly important in determining CO_2 emissions into the atmosphere. More mechanistic evaluation and future prospects have been progressing depending on the development of new experimental and modeling techniques (e.g., *in situ* leaf gas exchange measurements, eddy correlation and remote sensing) at several spatial and temporal scales (Saigusa et al. 2002; Sasai et al. 2007; Sato et al. 2007; Yamaji et al. 2008). Especially, among the approaches for understanding the roles of vegetation in carbon cycle mechanisms, models of ecosystem processes are essential tools (Farquhar et al. 1980; Leuning et al. 1995; de Pury and Farquhar 1997; Ito and Oikawa 2002). Models employing various canopy integration schemes are often used to derive vegetation canopy from leaf-scale information.

Typical inputs for canopy photosynthesis model include canopy architecture information, light distribution, and leaf or canopy photosynthetic capacity. Monsi and Saeki (1953) have developed the relationship between light interception and photosynthesis by accounting for canopy architecture of herbaceous vegetation. Their pioneering work (the M-S model) assumes randomness of foliage distribution and an exponential decrease in light intensity with cumulative leaf area towards the ground using a function based on the Beer-Lambert law. The M-S model has provided more insights into leaf biophysical (e.g., leaf structure and thickness) and biochemical (e.g., chlorophyll and other pigments, nitrogen) characteristics as well as better understanding of the processes that govern canopy photosynthesis (Terashima and Hikosaka 1995; Anten 1997; Hikosaka et al. 2001). However, for accurate calculations of canopy photosynthesis, non-randomness of foliage distribution must be taken into account, because leaf position and orientation directly affect light absorption (Takenaka 1994; Ackerly and Bazzaz 1995; Valladares and Pearcy 1998, 1999; Muraoka et al. 2003). For example, a vertical leaf orientation reduces light absorption at the surface layer of the canopy, but increases light penetration into deeper layers. Therefore, the spatial arrangement of leaves may have a greater effect on light environment within the canopy than the quantity of foliage itself.

Some studies have achieved model modifications by using representative leaf angles and considering two classes of leaves, sunlit and shaded leaves, in each layer separately (Norman 1980; Baldocchi and Harley 1995; de Pury and Farquhar 1997). However, detailed descriptions of canopy architecture and light distribution within a canopy are not sufficient to understand the total processes of canopy photosynthesis, because incoming solar radiation that reaches the top of the canopy is affected a lot already by many factors, including forest structure, species composition, slope, aspect, latitude, and climate of site through time. Especially in a deciduous forest, upper forest canopies strongly influence understory light environment both daily (Chazdon 1988; Pearcy et al. 1994) and seasonally (Baldocchi et al. 1984). Although most canopy photosynthesis models at forest stand-level do not consider understory vegetations, they play an extremely important role in the carbon sink/source function (Goulden et al. 1997; Sakai et al. 2006). A canopy photosynthesis model for understory vegetations should include detailed light simulation above and within the canopy, however the consequences for canopy photosynthesis are largely unknown.

In this study, we examine how and to what extent the different simulations of light distribution affect canopy photosynthesis estimations of an understory herbaceous vegetation, i.e., a dwarf bamboo, *Sasa senanensis* (Fr. Et Sav.) Rehder. As it is difficult to obtain representative and accurate measurements of canopy photosynthesis of understory to calibrate models, three different models are used. The quantitative, comparative studies are important to gain a better understanding of physiological process. Our effort is aimed at determining if there is the systematic bias in estimation of canopy photosynthesis, and if such the bias originates from the light simulation model.

MATERIALS AND METHODS

Study Site

The study site is located in the experimental forest of the River Basin Research Center of Gifu University, on the south-western slope of Mt. Norikura, Japan (36° 8' N, 137° 25' E, 1420m a.s.l.). Annual mean temperature and precipitation are 7.0 °C (daily maximum 25.3 °C; daily minimum -14.9 °C) and 2344 mm in 2001, respectively. Snow cover on the forest floor is usually from late November to mid April with maximum depth more than 2 m. The site is covered with a cool-temperate deciduous broadleaved secondary forest (approximately 50 years old), the forest canopy is approximately 15 - 20 m in height, and dominated by *Quercus crispula* Blume, *Betula platyphylla* Sukatchev var. *japonica* Hara and *Betula ermanii* Cham., and sub-canopy and shrub layers are dominated by *Acer rufinerve* Sieb. et Zucc., *Acer distylum* Sieb. et Zucc. and *Hydrangea paniculata* Sieb. Some evergreen coniferous trees, *Abies homolepis* Sieb. et Zucc., *Pinus parviflora* Sieb. et Zucc., and *Chamaecyparis pisifera* Endl., are distributed in very low densities. The understory vegetation is dominated by dwarf bamboo, *Sasa senanensis* (Fr. Et Sav.) Rehder. *Sasa* species constitute frequently almost pure stands with evergreen leaves in grassland and understory of deciduous broadleaved forest and sometimes in subalpine coniferous forest in eastern Asia (Nishimura et al. 2004).

Measurement of Leaf Gas Exchange

Measurements of leaf gas exchange were conducted under different forest canopy conditions (a matured forest canopy on 5 August 2001, an intermediate canopy on 23 October 2001, and after leaf shedding on 20 November 2001). Photosynthetic light response curves were measured with a portable photosynthesis measuring system (LI-6400, Li-Cor, USA). For all measurements, temperature, CO_2 concentration and leaf to air vapor pressure deficit (VPD) of the air entering the leaf chamber were set at the ambient air temperature (5 August; 18°C, 23 October; 10°C, 25 November; 5°C), 370 µmol mol^{-1} and < 1.0 kPa, respectively. Light was supplied by the red-blue LED lamp system of the LI-6400. The relationship between photosynthetically active photon flux density (PPFD, µmol m^{-2} s^{-1}) and photosynthetic rate was curve-fitted using the equation of Thornley (1976):

$$A = \frac{\alpha I + A_{g\,\max} - \sqrt{(\alpha I + A_{g\,\max})^2 - 4\theta\alpha I A_{g\,\max}}}{2\theta} - R \tag{1}$$

where A is the net photosynthetic rate, α is the initial slope of the light-response curve (i.e., apparent quantum yield), I is PPFD, $A_{g\,\max}$ is the light-saturated rate of gross photosynthesis, θ is the convexity of the light-response curve and R is the dark respiration rate. The apparent quantum yield was calculated by the linear-regression of the data obtained under 50 µmol m^{-2} s^{-1} of PPFD, while the light compensation point (LCP) was calculated as the x-intercept of this linear regression. Constant θ of 0.80 was given to all leaves.

Measurement of Stand Architecture

Six 80 cm x 80 cm plots were selected with different stand densities of *Sasa* along the whole range encountered in the forest on 5 October 2001. The position and orientation of individual leaves were measured under electromagnetic fields using a 3-D digitizing system, Polhemus 3 Space Isotrac II (Polhemus, USA). The specified accuracy of the system is 0.24 cm root mean square error (RMSE) for the x, y, or z positions, and 0.75° RMSE for the azimuth (Polhemus 1993). Operation with distances up to 152.4 cm is possible with reduced accuracy. All leaves were digitized for the six plots. Four points were digitized at the edge of each leaf, those are the base, the tip and the side edges. Then, four digitized points were treated as vectors from the base to determine the azimuth and inclination angles of the lamina and midrib. Leaf area was calculated by assuming it to be proportional to the square value of the midrib length. A representative leaf shape was traced on graph paper as an array of (x, y) coordinates. *Sasa* leaves were described as flat polygons with 16 points. The geometrical architecture of *Sasa* can be described by spatial distribution of assimilatory organs (mainly leaves). We consider here only significance of leaf distribution, and ignore the shading effect on culms and petioles. In addition, the assumption of constant stand architecture of *Sasa* is made from August to November, to keep the problem tractable. As foliation compensates for the shedding of old leaves, total leaf area and stand architecture of *Sasa* are kept almost constant throughout the year (Oshima 1961).

Calculations of Light Distribution and Photosynthesis

Light environment above the *Sasa* stand was evaluated by three indices, relative PPFD (RPPFD, ranging from $0 - 1$), the indirect site factor (ISF, ranging from $0 - 1$) and the direct site factor (DSF, ranging from $0 - 1$) under diffuse light conditions. RPPFD is the ratio of incident light above the *Sasa* stand to above the forest canopy with quantum sensors (Li-190SA, Li-Cor, USA). ISF and DSF are fractions of diffuse and direct light transmitted through the forest canopy, respectively, those are calculated from a hemispherical photograph using a commercially available program, HemiView 2.1 (Delta-T Devices, UK). Hemispherical photographs were taken above the *Sasa* stand (1.7 m above the ground) with a leveled digital camera (CoolPix 990, Nikon, Japan) equipped with a fish-eye lens (FC-E8, Nikon, Japan).

Canopy photosynthesis needs to be calculated continuously for the course of a day to obtain the daily total. In this study, canopy photosynthesis was calculated using three models. We focus on the effects of (1) the diurnal pattern of incident light above the *Sasa* stand and (2) the spatial distribution of incident light within the *Sasa* stand to estimate canopy photosynthesis.

1. The Monsi-Saeki 1 Model (M-S$_1$)

In the M-S$_1$ model, incident light above the *Sasa* stand (I_1) is given by multiplying the diurnal course of incident light above the forest canopy (I_0) and RPPFD above the *Sasa* stand [I_0 x RPPFD]. The diurnal course of incident light above the forest canopy follows the sine curve:

$$I_0 = I_{noon} \sin\left(\pi \frac{t - t_1}{t_2 - t_1} \right) \quad (t_1 \le t \le t_2)$$

$$I_0 = 0 \qquad\qquad (0 \le t < t_1, t_2 < t \le 24) \qquad\qquad (2)$$

where I_{noon} is the noon maximum incident light above the forest canopy, and t_1 and t_2 are the sunrise and sunset times, respectively. Incident light was calculated at 1-minute intervals throughout the day.

Vertical profile of incident light within the *Sasa* stand was calculated using the Beer's law which assumes light reduction with foliage increment:

$$I_i = I_1 \exp(-KF) \qquad\qquad (3)$$

where F is the leaf area cumulated from the top of the *Sasa* stand per unit ground area in 10 cm intervals. I_i is PPFD at F (within the *Sasa* stand). K is the light-extinction coefficient. Because K reflects the stand characteristics such as leaf angle distribution (Monsi and Saeki 1953; Saeki 1960), K was measured at each plot.

The estimated incident light above and within the *Sasa* stand was used for the calculation of canopy photosynthesis at each layer with Eq. (1) and the leaf gas exchange parameters.

2. The Monsi-Saeki 2 Model (M-S$_2$)

In the M-S$_2$ model, the diurnal course of incident light above the *Sasa* stand was calculated by a hemispherical photographic method using HemiView 2.1. The gap fraction was determined for 160 specific sky sectors on the photograph. The fraction of diffuse and direct light components was calculated from both ISF and DSF under the standard overcast sky (SOC) condition, respectively. In SOC condition, diffuse light is three times brighter at the zenith angle than the horizon. The M-S$_2$ model also used the Beer's law to describe vertical profile of incident light within the *Sasa* stand.

3. The Y-Plant Model

The Y-plant model can account for irregularities in stand description such as leaf angle and sun's rays using hemispherical photographs and a 3-D digitizer, and calculate photosynthesis for individual leaves (Pearcy and Yang 1996; Valladares and Pearcy 1998, 1999; Valladares et al. 2002; Muraoka et al. 2003). The details of the model and architecture inputs can be found in the paper of Pearcy and Yang (1996). Leaf absorbance and transmittance were assumed 0.85 and 0.10, respectively. The outputs of incident light and photosynthesis were then integrated for each layer to compare with the outputs of the former two M-S models.

Simulation results of canopy photosynthesis were compared under completely clear or cloudy sky conditions. Under completely cloudy sky conditions, it was assumed that all light was 100% diffuse and the total amount of incident light was adjusted to 30% of a clear day. The diurnal patterns of incident light above the forest canopy were the same among three models. The day length and the noon maximum incident light, which were calculated by time of day and latitude in the Y-plant model, were applied in three models. The day lengths were 816, 644 and 594 minutes, and the noon maximal incident lights were 2036, 1393, 1133 μmol m^{-2} s^{-1} for 5 August, 23 October and 20 November 2001, respectively. To generalize the gap effects by a hemispherical photographic method (the M-S$_2$ and Y-plant models), calculations were made using all photographs taken for the other plots. Thus there are 36 samples (6 plots x 6 photographs) for each season. We consider here only significance of light and other factors, such as air temperature, water vapor partial pressure and CO_2 concentration, are assumed to be constant over the canopy depth although vertical gradients of these factors exist in most plant canopies. The differences in the simulation results are therefore entirely due to the different calculation of incident light above and within the *Sasa* stands.

RESULTS AND DISCUSSION

Stand architecture and light environment

Table 1 shows the stand architectural characteristics and light environmental factors at the six plots of *Sasa*. LAI ranged from 0.86 to 2.72 (1.81 ± 0.71, mean ± SD) and leaf numbers ranged from 76 to 200 (129 ± 41).

Table 1. Summary of stand architecture and light environmental factors for six plots of *Sasa*

		Plot 1	Plot 2	Plot 3	Plot 4	Plot 5	Plot 6	Mean ± SD
LAI (m m^{-2})		0.86	0.99	1.74	1.95	2.60	2.72	1.81 ± 0.71
Plant height (cm)		84.0	115.0	162.8	136.0	140.0	165.0	133.8 ± 28.0
Leaf number		76	84	129	122	200	164	129 ± 41
Leaf angle (°)		30.3±19.3	32.0±21.7	36.0±18.5	33.2±18.5	28.5±16.4	28.8±14.8	31.1 ± 17.9
K (light-extinctioin coefficient)		1.06	1.23	1.00	0.68	0.56	0.58	0.85 ± 0.27
Relative PPFD	7 Aug.	0.045	0.043	0.049	0.066	0.059	0.069	0.055 ± 0.011
	23 Oct.	0.199	0.177	0.170	0.183	0.223	0.220	0.195 ± 0.022
	20 Nov.	0.435	0.434	0.452	0.520	0.489	0.482	0.469 ± 0.034
ISF (indirect site factor)	7 Aug.	0.098	0.100	0.105	0.100	0.111	0.098	0.102 ± 0.005
	23 Oct.	0.219	0.202	0.195	0.186	0.211	0.213	0.204 ± 0.012
	20 Nov.	0.478	0.474	0.562	0.477	0.552	0.535	0.513 ± 0.041
DSF (direct site factor)	7 Aug.	0.068	0.068	0.107	0.126	0.100	0.119	0.098 ± 0.025
	23 Oct.	0.202	0.206	0.100	0.203	0.150	0.238	0.183 ± 0.050
	20 Nov.	0.441	0.432	0.571	0.335	0.537	0.433	0.458 ± 0.084

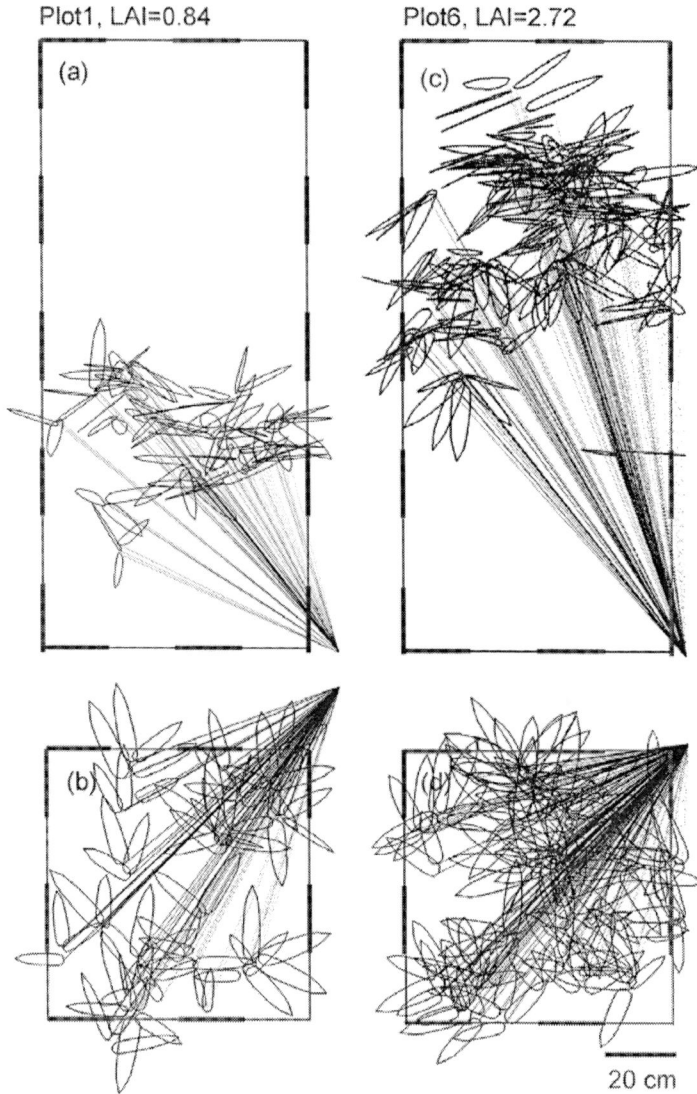

Figure 1. Stand architecture of *Sasa* reconstructed by using the 3-D digitizer and the Y-plant model for the lowest- (a, b) and highest- (c, d) density stands. Views from above (a, c) and the side (b, d) of the stands.

Leaf surface angle from a horizontal plane was $31.1 \pm 17.9°$. Positive correlations existed between LAI and leaf number ($r = 0.944$) and stand height ($r = 0.788$), and a negative correlation existed between LAI and K ($r = -0.930$). Figure 1 shows the sample architectures of the lowest- (Plot 1) and highest- (Plot 6) density stands as reconstructed by the Y-plant model, and Figure 2 shows the vertical profiles of leaf area density (LAD) and RPPFD in 10 cm intervals at the same plots in Figure 1. The summation of the leaf area by the 3-D digitizer gave the vertical profile of LAD and RPPFD for each layer. *Sasa* stands were characterized by the development of leaf area at high positions on the culms (Figs. 1 and 2). Younger leaves were produced in upper positions and older leaves occupied lower and less illuminated positions (Figure 2).

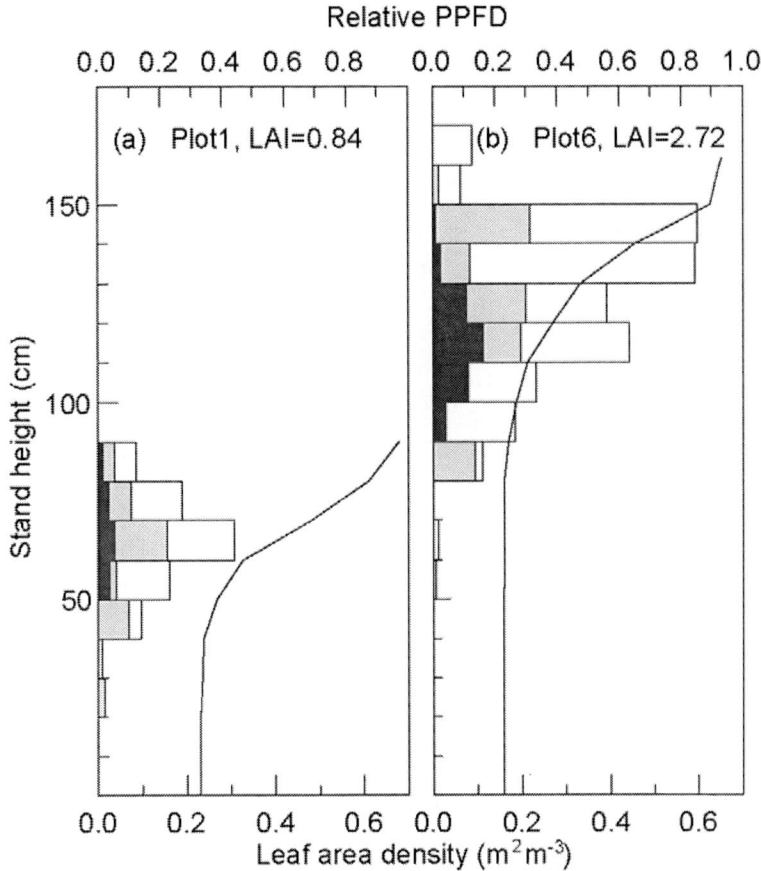

Figure 2. Vertical profiles of leaf area density of different leaf age; current- (white bar), 1- (gray bar) and 2- (solid bar) year old, and relative PPFD for the lowest- (a) and highest- (b) density stands of *Sasa*. Leaf area distribution was converted from the 3-D measurements (see Figure 1).

Figure 3 shows samples of the hemispherical photographs taken under different forest canopy conditions. Forest canopy structure that associated with leaf phenology varied the amount of light. Although three light environmental factors (RPPFD, ISF and DSF) varied little among the six plots, their seasonal changes were remarkable (Table 1). RPPFD changed from 0.055 ± 0.011 on August to 0.469 ± 0.034 on November above the *Sasa* stand. Therefore detailed understanding canopy photosynthesis of the understory vegetations requires long-term estimates (seasonally) as well as readily instantaneous estimates (daily) in the deciduous forest. Ideally, incident light should be measured continuously throughout the growing season in order to sample the spatial and temporal complexity of the light environment.

Leaf Gas Exchange Characteristics

Younger leaves had significantly higher light-saturated photosynthetic rates (A_{max}) (Table 2).

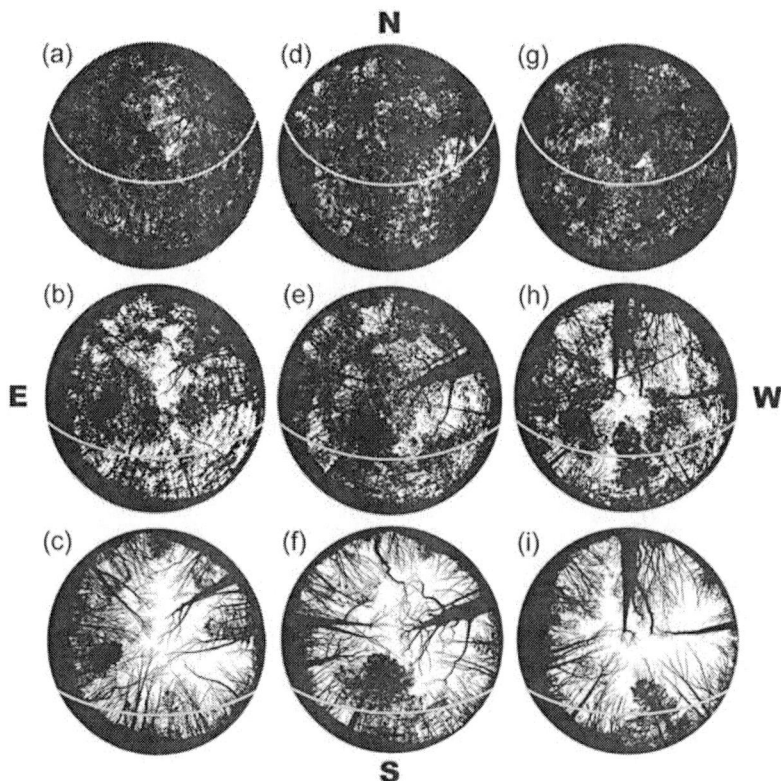

Figure 3 Examples of hemispherical photograph for three plots (plot1 (a, b, c), plot 3 (d, e, f) and plot 5 (g, h, i)) taken under different forest canopy conditions; a mature forest on 5 August (a, d, g), an intermediate canopy on 23 October (b, e, h) and after leaf shedding on 20 November 2001 (c, f, i). The curved line shows the sun orbit.

This can be explained by leaf senescense, however higher A_{max} might be occurred in higher position where light availability is higher (Figure 2). In many herbaceous plants, there is evidence that the relatively high A_{max} can be a result of acclimation of the species to the understory environment (Field 1983; Hikosaka et al. 1994; Hirose et al. 1997). Similarly, apparent quantum yield (α) and dark respiration rate (R) of younger leaves tended to be higher than older ones, although differences were not significant.

A_{max} was also related to environmental conditions throughout the season (Table 2). The leafless periods of forest canopy are probably important for the annual carbon budget of understory vegetation (Sakai et al. 2006). In order to take full advantage of these periods of high light environment, *Sasa* had high photosynthetic characteristics even at relatively low temperatures on 23 October. However, A_{max} of all ages decreased approximately by half on 25 November because the air temperature was close to freezing.

Photosynthetic rates saturated at relatively low light intensity, ranging from approximately 150 to 250 μmol m^{-2} s^{-1}, and light compensation points (LCP) were very low, ranging from 2 to 4 μmol m^{-2} s^{-1}. *Sasa* could be very flexible to adapt to light environment of forest understory, which is quite important. Species that can modify photosynthetic characteristics in response to environmental conditions would have a better chance of surviving understory conditions.

Table 2. Leaf photosynthetic characteristics for current-, 1- and 2-year-old leaves of *Sasa* throughout the season

	Current-year			1-year-old			2-year-old		
	5 Aug.	23 Oct.	25 Nov.	5 Aug.	23 Oct.	25 Nov.	5 Aug.	23 Oct.	25 Nov.
A_{max} (μmol m^{-2} s^{-1})	7.66±1.01	8.88±1.07	4.46±1.16	6.37±1.05	6.24±1.57	3.71±0.68	4.74±1.07	4.49±1.24	2.81
R (μmol m^{-2} s^{-1})	0.13 ± 0.03	0.13±0.07	0.13±0.02	0.09±0.02	0.12±0.06	0.13±0.05	0.09±0.02	0.11±0.09	0.10
α (μmol CO$_2$ μmol^{-1} photons)	0.077±0.007	0.086±0.006	0.059±0.023	0.068±0.009	0.085±0.008	0.047±0.009	0.059±0.008	0.086±0.006	0.04
	(n=7)	(n=6)	(n=3)	(n=7)	(n=6)	(n=3)	(n=6)	(n=6)	(n=2)

Daily Light Capture and Canopy Photosynthesis

Figure 4 shows the daily absorbed photosynthetically active radiation ($APAR_{day}$, mol m^{-2} day^{-1}) and daytime net canopy photosynthesis (A_{day}, mmol m^{-2} day^{-1}) of the stand per unit ground area using the M-S_1, M-S_2 and Y-plant models under different forest canopy conditions on 5 August, 23 October and 20 November 2001. $APAR_{day}$ did not statistically differ much among the models under both clear and cloudy sky conditions, although $APAR_{day}$ in the M-S_2 and Y-plant models using the hemispherical photographic method showed lower value than that in the M-S_1 model on 20 November. Forest canopy was after leaf shedding on 20 November, but incident light from the sun was almost intercepted by trunks and branches of upper forest canopies because of the low solar elevation in the hemispherical photographic method (Figure 3).

A_{day} varied greatly among the models in order of the M-S_1, M-S_2 and Y-plant models under clear sky conditions (Tukey-Kramer, $p<0.01$, $n=36$). The M-S models based on the Beer's law gave higher A_{day} than the Y-plant model that considered more realistic light distribution patterns. A_{day} in the M-S_1 and M-S_2 models were the average 213% and 78% higher than that in the Y-plant model, respectively. On the other hand, although the significant difference of A_{day} was observed between the M-S_1 and Y-plant models (Student's t-test, $p<0.05$, $n=6$), overestimation of A_{day} by the M-S_1 model reduced from 213% to 47% under cloudy sky conditions. These results suggest that canopy photosynthesis model can be simplified if cloudy sky conditions were assumed in the analysis. The significant overestimation of A_{day} as a result of ignoring dynamic responses under clear sky conditions can be attributed to the different manner of calculating (i) the diurnal pattern of incident light above the stands and (ii) the light distribution pattern within the stands, as follows.

Effect of Incident Light above the Stand

In order to demonstrate the effect of diurnal pattern of incident light above the *Sasa* stand for estimating A_{day}, the simulation results from the M-S_1 and M-S_2 models were compared. In the M-S_1 and M-S_2 models, the different approaches were used to calculate the diurnal pattern of incident light above the *Sasa* stand following the sine-curved method and the hemispherical photographic method, respectively, and the vertical profiles of incident light within the *Sasa* stand were calculated following the same approach through Beer's law. Figure 5 shows the frequency distributions of incident light above the *Sasa* stand and the photosynthetic rate of current-year leaf on 23 October 2001 corresponding to each incident light class. Frequency distributions of incident light above the *Sasa* stand were remarkably different between the models (the sine-curved method vs the hemispherical photographic method). In the M-S_1 model, the frequency distributions were aggregative with relatively high light intensity (PPFD: 20 – 200 μmol m^{-2} s^{-1}, 7 August 2001). Under these light conditions, photosynthetic rates kept relatively high throughout the day, because photosynthetic rate reached saturation at light intensity of approximately 200 μmol m^{-2} s^{-1}. On the other hand, the frequency distributions in the M-S_2 model (the hemispherical photographic method) were disaggregative. The light environment above the *Sasa* stand was very low (< 20 μmol m^{-2} s^{-1}, 7 August 2001) during most of the day, and the *Sasa* stand could receive high light intensity (> 400 μmol m^{-2} s^{-1}) for less than 2 h, hence the photosynthetic rates kept low.

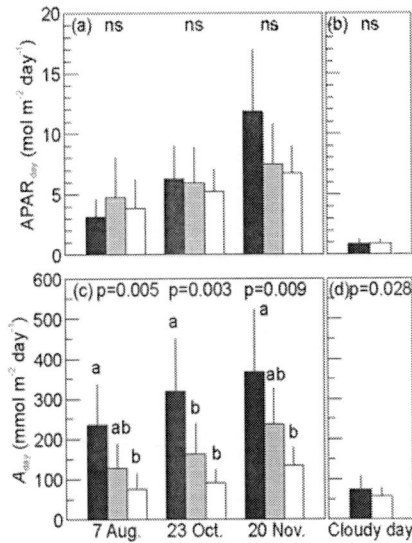

Figure 4. Daily light capture (a, b) and daytime photosynthesis (c, d) per unit ground area of *Sasa* under different forest canopy conditions. Daily light capture and daytime photosynthesis were calculated using three models under clear sky conditions (a, c) and cloudy sky conditions (b, d); the M-S$_1$ (solid bars), M-S$_2$ (gray bars) and Y-plant (white bars) models. Mean ± SD are plotted. Different letters indicate the significant difference determined by Tukey-Kramer test (a, c)(n=36) and Student's t-test (b, d)(n=6).

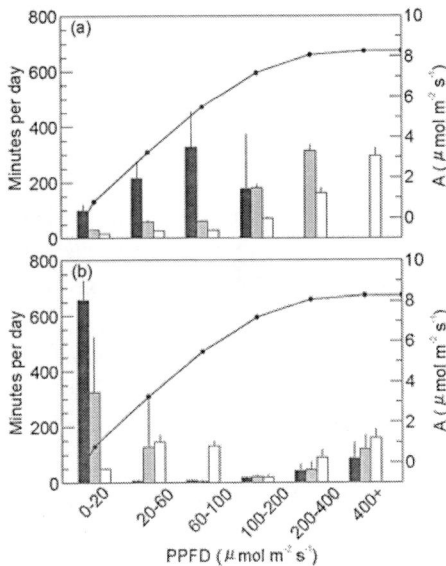

Figure 5 Frequency distribution of incident light above the *Sasa* stand. Incident light was simulated by the sine-curved method (a) and the hemispherical photographic method (b) on 5 August (solid bars), 23 October (gray bars) and 20 November 2001 (white bars). Mean ± SD for the six plots are plotted. Single leaf photosynthetic rate of current-year on 23 October 2001 corresponding to each incident light class are also plotted.

In consequence, although $APAR_{day}$ did not vary significantly between the M-S_1 and M-S_2 models to the exclusion on 23 October, A_{day} in the M-S_1 model was overestimated by 76% in comparison to that in the M-S_2 model (Figure 4).

Many studies report that overestimations of A_{day} is caused by interactions between the non-linearity of light-photosynthesis curve and the diurnal fluctuations in incident light (Norman 1980; Sims and Pearcy 1993; Leuning et al. 1995; de Pury and Farquhar 1997; Sakai et al. 2005). The sinc-curved method (the M-S_1 model) could not mimic the actual understory light environment that was characterized by mostly strong sunflecks with intermittent short period. On the other hand, the hemispherical photographic method (the M-S_2 model) could not only simulate the appropriate variability of diurnal pattern, but also simulate the right magnitude of light intensity (Canham 1988; Canham et al. 1990; Rich et al. 1993). There is no doubt that the detailed estimation of diurnal pattern of incident light is one of the determinant factors for the accurate photosynthesis model. Canopy photosynthesis model certainly requires instantaneously diurnal distribution pattern.

Effect of Light Distribution within the Stand

Differences in A_{day} ware also caused by differences in light distribution pattern within the *Sasa* stand. We compared the different simulations to reveal the effect of the light distribution pattern within the *Sasa* stand for estimating A_{day} (the M-S_2 model vs the Y-plant model). The incident light above the *Sasa* stand is the same pattern between M-S_2 and Y-plant models (calculated by hemispherical photographs), and the differences are explained by the effects of foliage clustering. In the M-S_2 model, the horizontally homogeneous leaf distribution is assumed and the vertical profile of light environment exponentially decreases with cumulative leaf area. However, the actual instantaneous profile of incident light does not. In Y-plant model, leaf position and orientation are taken into consideration to evaluate light absorption and photosynthesis.

Figure 6 shows the samples of the vertical profile of $APAR_{day}$ and A_{day} per single leaf (whose area is the mean value of leaves in each layer) in the M-S_2 and Y-plant models on 7 August 2001. Each layer contained 0 – 45 leaves depending on its height. Light absorption per single leaf was highly variable in the Y-plant model (see the SD on Figure 6). Although the leaf dispersion allowed more light penetration within the stand, slight changes in the angle between a perpendicular to the surface and incident light could result in big changes in light absorption. In addition, leaf overlap also significantly affected the light condition of leaf surface. Figure 7 shows the seasonal changes in the effects of LAI (stand density) on $APAR_{day}$ and A_{day} per whole stand in three models. Although $APAR_{day}$ in the two M-S models increased with increasing LAI, $APAR_{day}$ in the Y-plant model tended to saturate with increasing LAI (> approximately 2.0). Although minimum leaf overlaps were achieved to increase more light capture per unit leaf area at lower LAI (Figure 1b), higher LAI that attained full canopy closure led more leaf overlaps (Figure 1d). These leaf distributions generate complicated non-stationary patterns of direct, diffuse and scattered light. The uppermost leaves could absorb high incident light such as sunflecks, but other leaves absorbed diffuse or scattered lights with less intensity.

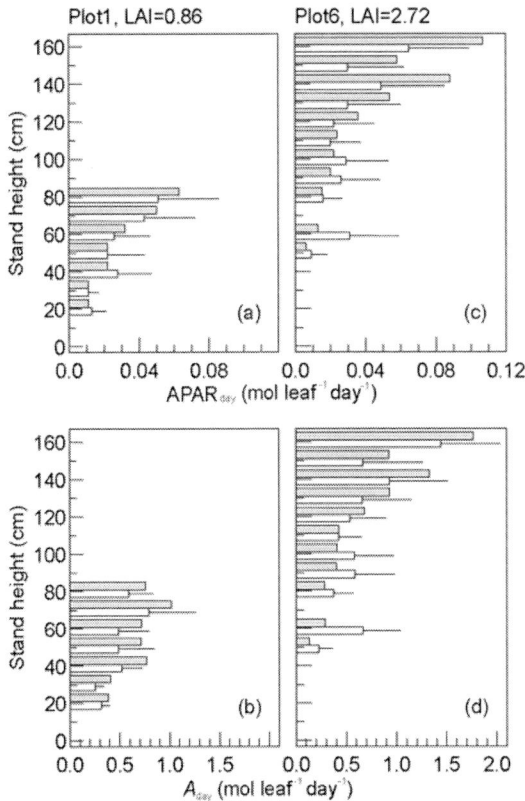

Figure 6. Vertical distribution of daily light capture (a, c) and daytime photosynthesis (b, d) per single leaf simulated by the M-S_2 (gray bars) and Y-plant (white bars) models for the lowest- (a, b) and highest- (c, d) density stands. Mean ± SD are plotted. Each layer includes 0 - 45 leaves depending on its height.

The effects of clumping by leaf overlaps were found as LAI above 2. Reflecting the clumping effect such as leaf overlaps, A_{day} with lower LAI were found to be relatively close among the models, but the differences increased with increasing LAI. This result is supported by a modeling exercise by de Pury and Farquhar (1997), where a big-leaf model was compared to multi-layered and sun/shade leaf models. Their result indicates that detailed light characterization between sunlit and shaded leaves further reduced the errors associated with simplification of the canopy architecture. The Y-plant model detected the large fraction of the shaded leaves at given time and position, and showed the large differences in light absorption between sunlit and shaded leaves. If ignored, these large temporal and spatial variations in light could cause large errors in canopy photosynthesis calculations. Consequently, different vertical distributions of light environment created differences (78 %) in A_{day} between the M-S_2 and Y-plant models.

Assessment of Canopy Photosynthesis Model

Application of canopy photosynthesis models is not possible without describing the stand architecture of the vegetation and its effect on light absorption by the leaves in detailed.

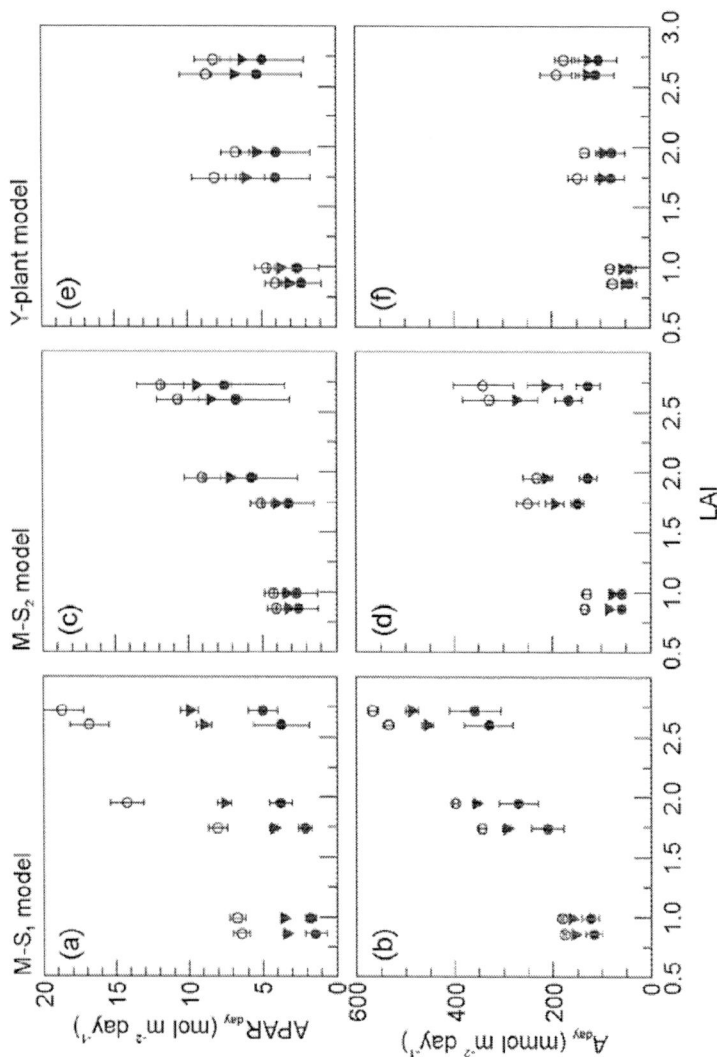

Figure 7. Effects of LAI on the simulated daily light capture (a, c, e) and daytime photosynthesis (b, d, f) per unit ground area by the M-S$_1$ (left), M-S$_2$ (intermediate) and Y-plant (right) models under different forest canopy conditions. Closed circle, triangle and open circle symbols indicate simulations on 5 August, 23 October and 20 November 2001, respectively. The data are presented as the mean ± SD.

We used three models (the M-S$_1$, M-S$_2$ and Y-plant models) to assess how different light simulations affect canopy photosynthesis. Specific attentions were paid to the effects of light environment above and within the *Sasa* stand under different environment conditions throughout the season.

The M-S$_1$ model was the most simple and rapid data processing, therefore this simple empirical model may lack a predictive capacity. The magnitude of A_{day} in the M-S$_1$ model was too large compared with values in the Y-plant model under clear-sky conditions (Figure 4). The Y-plant model had the most advantage for providing realistic representation of stand architecture and light environment among three models examined in this study. In principle, all information concerning spatio-temporal properties of light is valuable, and any averaging should be avoided in the light simulation (Larocque 2002). However, the Y-plant model was much time consuming to obtain detailed stand architecture information, and thus limited the number of plots that can be sampled in the field. Although substantial improvements in canopy photosynthesis are obtainable by suitably integrating parameters, the regional extrapolation from the limited plot data is a challenging task because of the large spatial heterogeneity and temporal dynamics of forest ecosystems (Sakai and Akiyama 2005). When different scales are involved, one must consider sufficiently simple but robust canopy photosynthesis models. The relatively close match of A_{day} between the M-S$_2$ and Y-plant model may suggest that simpler models based on Beer's law with a few parameters (Eq. 3) are reasonable for estimation of canopy photosynthesis to the extent possible, although at least sunlit and shaded leaves should be calculated.

It is necessary to understand the possibilities and limitations of the proposed model. If there is less difference in accuracy among the models, then the criterion can be used for selecting a simple model of function minimization. Our study showed the possible ranges how much A_{day} was affected by light distribution patterns, stand density and seasonal changes in the light environment. However, our understanding of responses of canopy photosynthesis to climate change is still incomplete, because all models are only approximations. A remaining question is how closely they relate to measured data of canopy photosynthesis. In the future works, we need to compare the estimations with measured data. With further validation and development, the canopy photosynthesis model could have the potential to be applied at large spatial scales, and would be used to bridge the gap between local measurements and regional estimations.

CONCLUSION

To demonstrate the importance of the light distribution pattern and the stand architecture for canopy photosynthesis, the outputs of the three models (the M-S$_1$, M-S$_2$ and Y-plant models) were compared. (i) The methods to determine the light environments above the *Sasa* stand were critical for canopy photosynthesis. The use of diurnal pattern of incident light estimated by a sine curve led to a 76% overestimation of A_{day} in the model separating sunflecks and diffuse light (i.e., the M-S$_1$ model vs the M-S$_2$ model). (ii) Estimations of canopy photosynthesis were also affected by light distributions within the stand. Less calculation of the clumping effect such as leaf overlaps caused a 78% overestimation of A_{day}, and the error was larger with higher LAI (i.e., the M-S$_2$ model vs the Y-plant model). Consequently, A_{day} could be overestimated by an average of 213% as a result of ignoring dynamic responses of light. These results suggest that more detailed treatments of light distribution pattern are needed for overcoming some of the errors associated with simplifying assumptions. The Y-plant model had the most advantage for providing realistic representation of stand architecture and light environment among three models examined in this study. However, this 3-D light transfer model is time consuming to parameterize geographical distribution of single leaf area and orientation, some of which are not easily obtained. Therefore, for large scale estimation, the light distribution need to be computed with much simpler models based on a few parameters such as Beer's law (Eq. 3). It is necessary to understand the possibilities and limitations of the proposed model and to determine which models make more efficient use of survey data. The quantitative, comparative studies are important to gain a better understanding of physiological process. Our study showed the possible ranges how much A_{day} was affected by light distribution patterns, stand density and seasonal changes in the light environment.

REFERENCE

Ackerly D. D. and Bazzaz F. A. (1995) Seedling crown orientation and interception of diffuse radiation in tropical forest gaps. *Ecology*, *76*, 1134-1146.

Anten N. P. R. (1997) Modelling canopy photosynthesis using parameters determined from simple non-destructive measurements. *Ecological Research*, *12*, 77-88.

Baldocchi D. D. and Harley P. C. (1995) Scaling carbon dioxide and water vapour exchange from leaf to canopy in deciduous forest 2. Model testing and application. *Plant, Cell and Environment*, *18*, 1157-1173.

Baldocchi D. D., Hutchison B., Matt D. and McMillen R. (1984) Seasonal variations in the radiation regime within an oak-hickory forest. *Agricultural and Forest Meteorology*, *33*, 177-191.

Canham C. D. (1988) An index for understory light levels in and around canopy gaps. *Ecology*, *69*, 1634-1638.

Canham C. D., Denslow J. S., Platt W. J., Runkle J. R., Spies T. A. and White P. S. (1990) Light regimes beneath closed canopies and tree-fall gaps in temperate and tropical forests. *Canadian Journal of Forest Research*, *20*, 620-631.

Chazdon R. L. (1988) Sunflecks and their importance to forest understory plants. *Advances in Ecological Research*, *18*, 1-63.

de Pury D. G. G. and Farquhar G. D. (1997) Simple scaling of photosynthesis from leaves to canopies without the errors of big-leaf models. *Plant, Cell and Environment*, *20*, 537-557.

Farquhar G. D., von Caemmerer S. and Berry J. A. (1980) A biochemical model of photosynthetic CO_2 assimilation in leaves of C3 species. *Planta*, *149*, 78-90.

Field C. (1983) Allocating leaf nitrogen for the maximization of carbon gain: leaf age as a control on the allocation program. *Oecologia*, *56*, 341-347.

Goulden M. L. and Crill P. M. (1997) Automated measurements of CO_2 exchange at the moss surface of a black spruce forest. *Tree Physiology*, *17*, 537-542.

Hikosaka K., Nagashima H., Harada Y. and Hirose T. (2001) A simple formulation of interaction between individuals competing for light in a monospecific stand. *Functional Ecology*, *15*, 642-646.

Hikosaka K., Terashima I. and Katoh S. (1994) Effects of leaf age, nitrogen nutrition and photon flux density on the distribution of nitrogen among leaves of a vine (*Ipomoea tricolor* Cav.) grown horizontally to avoid mutual shading of leaves. *Oecologia*, *97*, 451-457.

Hirose T., Ackerly D. D., Traw M. B., Ramseier D. and Bazzaz F. A. (1997) CO_2 elevation, canopy photosynthesis, and optimal leaf area index. *Ecology*, *78*, 2339-2350.

Ito A. and Oikawa T. (2002) A simulation model of the carbon cycle in land ecosystems (Sim-CYCLE): A description based on dry-matter production theory and plot-scale validation. *Ecological Modelling, 151*, 143-176.

Larocque G. R. (2002) Coupling a detailed photosynthetic model with foliage distribution and light attenuation functions to compute daily gross photosynthesis in sugar maple (*Acer saccharum* Marsh.) stands. *Ecological Modelling, 148*, 213-232.

Leuning R., Kelliher F. M., De Pury D. G. G. and Schulze E.-D. (1995) Leaf nitrogen, photosynthesis, conductance and transpiration: Scaling from leaves to canopies. *Plant, Cell and Environment, 18*, 1183-1200.

Monsi M. and Saeki T. (1953) Über den Lichtfaktor in den Pflanzengesellschaften und seine Bedeutung für die Stoffproduktion. *Japanese Journal of Botany, 14*, 22-52.

Muraoka H., Koizumi H. and Pearcy R. W. (2003) Leaf display and photosynthesis of tree seedlings in a cool-temperate deciduous broadleaf forest understorey. *Oecologia, 135*, 500-509.

Nishimura N., Matsui Y., Ueyama T., Mo W., Saijo Y., Tsuida S., Yamamoto S. and Koizumi H. (2004) Evaluation of carbon budgets of a forest floor *Sasa senanensis* community in a cool-temperate forest ecosystem, central Japan. *Japanese Journal of Ecology, 54*, 143-158.

Norman J. M. (1980) Interfacing leaf and canopy irradiance interception models. CRC Press, Florida.

Oshima Y. (1961) Ecological studies of *Sasa* communities II. Seasonal variations of productive structure and annual net production in *Sasa* communities. *Botanical Magazine of Tokyo, 74*, 280-290.

Pearcy R. W., Chazdon R. L., Gross L. J. and Mott K. A. (1994) Photosynthetic utilization of sunflecks: A temporally patchy resource on a time scale of seconds to minutes. In: Exploitation of Environmental Heterogeneity by Plants (eds. M. M. Caldwell, R. W. Pearcy) 175-208. Academic Press, New York.

Pearcy R. W. and Yang W. M. (1996) A three-dimensional crown architecture model for assessment of light capture and carbon gain by understory plants. *Oecologia, 108*, 1-12.

Polhemus (1993) 3SPACE ISOTRAK II User's Manual. Polhemus, Vermont.

Rich P. M., Clark D. B., Clark D. A. and Oberbauer S. F. (1993) Longterm study of solar radiation regimes in a tropical wet forest using quantum sensors and hemispherical photography. *Agricultural and Forest Meteorology, 65*, 107-127.

Saeki S. (1960) Interrelationships between leaf amount, light distribution and total photosynthesis in a plant community. *Botanical Magazine of Tokyo, 73*, 55-63.

Saigusa N., Yamamoto S., Murayama S., Kondo H. and Nishimura N. (2002) Gross primary production and net ecosystem exchange of a cool-temperate deciduous forest estimated by the eddy covariance method. *Agricultural and Forest Meteorology, 112*, 203-215.

Sakai T. and Akiyama T. (2005) Quantifying the spatio-temporal variability of net primary production of the understory species, *Sasa senanensis*, using multipoint measuring techniques. *Agricultural and Forest Meteorology, 134*, 60-69.

Sakai T., Saigusa N., Yamamoto S. and Akiyama T. (2005) Microsite variation in light availability and photosynthesis in a cool-temperate deciduous broadleaf forest in central Japan. *Ecological Research, 20*, 537-545.

Sakai T., Akiyama T., Saigusa N., Yamamoto S. and Yasuoka Y. (2006) The contribution of gross primary production of understory dwarf bamboo, *Sasa senanensis*, in a cool-temperate deciduous broadleaved forest in central Japan. *Forest Ecology and Management*, *236*, 259-267.

Sasai T., Okamoto K., Hiyama T. and Yamaguchi Y. (2007) Comparing terrestrial carbon fluxes from the scale of a flux tower to the global scale. *Ecological Modelling*, *208*, 135-144.

Sato H., Itoh A. and Kohyama T. (2007) SEIB-DGVM: A new Dynamic Global Vegetation Model using a spatially explicit individual-based approach. *Ecological Modelling*, *200*, 279-307.

Sims D. A. and Pearcy R. W. (1993) Sunfleck frequency and duration affects growth rate of the understorey plant, *Alocasia macrorrhiza*. Functional Ecology, 7, 683-689.

Takenaka A. (1994) Effects of leaf blade narrowness and petiole length on the light capture efficiency of a shoot. *Ecological Research*, *9*, 109-114.

Terashima I. and Hikosaka K. (1995) Comparative ecophysiology of leaf and canopy photosynthesis. *Plant, Cell and Environment*, *18*, 1111-1128.

Thornley J. H. M. (1976) Mathematical models in plant physiology. Academic Press, London.

Valladares F. and Pearcy R. W. (1998) The functional ecology of shoot architecture in sun and shade plants of *Heteromeles arbutifolia* M. Roem., a Californian chaparral shrub. *Oecologia*, 114, 1-10.

Valladares F. and Pearcy R. W. (1999) The geometry of light interception by shoots of *Heteromeles arbutifolia*: Morphological and physiological consequences for individual leaves. *Oecologia*, *121*, 171-182.

Valladares F., Skillman J. B. and Pearcy R. W. (2002) Convergence in light capture efficiencies among tropical forest understory plants with contrasting crown architectures: A case of morphological compensation. *American Journal of Botany*, *89*, 1275-1284.

Yamaji T., Sakai T., Endo T., Baruah P., Akiyama T., Saigusa N., Nakai Y., Kitamura K., Ishizuka M. and Yasuoka Y. (2008) Scaling-up technique for net ecosystem productivity of deciduous broadleaved forests in Japan using MODIS data. *Ecological Research*, *23*, 765-775.

In: Forest Canopies: Forest Production, Ecosystem…   ISBN 978-1-60741-457-5
Editor: J. D. Creighton and P. J. Roney   © 2009 Nova Science Publishers, Inc.

Chapter 4

LIDAR REMOTE SENSING FOR FOREST CANOPY STUDIES

A. Farid [a,1], D.C. Goodrich [b] and S. Sorooshian [c]

[a] Department of Water Engineering, Ferdowsi University of Mashhad, Iran
[b] USDA-ARS-SWRC, Southwest Watershed Research Center, Tucson, AZ, USA
[c] Department of Civil and Environmental Engineering, University of California,
Irvine, CA, USA

ABSTRACT

Remote sensing has facilitated extraordinary advances in modeling, mapping, and the understanding of ecosystems. Applications of remote sensing involve either images from passive optical systems, such as Aerial Photography and the Landsat Thematic Mapper, or, active Radar sensors such as RADARSAT. These types of remote sensors have proven to be satisfactory for many forest applications, such as mapping and classifying land cover into specific classes and, in some biomes, estimating aboveground biomass and Leaf Area Index (LAI). However, conventional sensors have significant limitations for ecological and forest applications. The sensitivity and accuracy of these devices have repeatedly been shown to fall with increasing aboveground biomass and LAI. They are also limited in their ability to represent the spatial patterns. They produce only two-dimensional (x and y) images, which cannot fully represent the three dimensional structure of the forest canopy. Ecologists have long understood that the presence of specific organisms and the overall richness of wildlife communities can be highly dependent on the three-dimensional spatial pattern of vegetation. Individual bird species, in particular, are often associated with specific three dimensional features in riparian forests. Additionally, aspects of forests, such as productivity, may be related to forest canopy structure.

Lidar (light detecting and ranging) is an alternative remote sensing technology that promises to both increase the accuracy of biophysical measurements and extend spatial analysis into the third dimension (z). Lidar sensors directly measure the three-dimensional distribution of forest canopies as well as sub-canopy topography, therefore

---

1 Ali Farid, Ferdowsi University of Mashhad, Tel.: +01198-915-517-6454; fax: +01198-511-761-0681. E-mail address: afaridh@yahoo.com.

providing high resolution topographic maps and highly accurate estimates of tree height, cover, and canopy structure. In addition, lidar has been shown to accurately estimate LAI and aboveground biomass, even in those high biomass ecosystems, where passive optical and active radar sensors typically fail to do so. Estimation of forest structural attributes, such as LAI, is an important step in identifying the amount of water use in forest areas.

INTRODUCTION

Forests are often described in terms of their composition and structure. While composition typically refers to the presence and abundance of (floristic) species, structure is more broadly defined as "the physical arrangement and characteristics of the forest" Ecological function is an expression of and is affected by canopy vertical structure. Knowledge of canopy vertical structure is particularly important in assessing the potential value of forest resources for the production of private (water quality/quantity, recreation, fisheries and wildlife habitat, carbon sequestration) and public (wood, fiber, no timber forest products) goods. Basic forest inventory procedures measure only a subset of the total composition and structure in a given area, as one might, for example, tally only trees of a certain species, of those that are larger than some threshold breast-height diameter. The resulting descriptors of forest composition and structure are essentially those that are (1) useful to forest managers and (2) relatively easy to obtain using basic field skills and tools.

For more demanding investigations and analyses, however, canopy vertical structure is typically measured using some combination of in situ and remotely sensed data. Remotely sensed data sources include digital photogrammetry, large footprint lidar sensors (Means et al. 1999), interferometric SAR, and small footprint lidar (Farid et al. 2008). Of these possibilities only digital photogrammetry and small footprint lidar data are both (i) widely available commercially, and (ii) suitable for structural analyses at the scale of management, which is the stand or sub stand unit on which silvicultural prescriptions (establishment, fertilization, thinning, release, and harvest) are made. Of these two data sources, small-footprint lidar data has become the most common.

LIDAR SENSORS

The basic measurement made by a lidar instrument is the distance between the sensor and a target surface, obtained by determining the elapsed time between the emission of a laser pulse and the arrival of the reflection of that pulse (the return signal) at the sensor's receiver. Multiplying this time interval by the speed of light results in a measurement of the round-trip distance traveled and dividing that figure by two, yields the distance between the sensor and the target (Bachman 1979). When the vertical distance between a sensor contained in a level-flying aircraft and the Earth's surface is repeatedly measured along transect the result is an outline of both the ground surface and any vegetation obscuring it. Even in areas with high vegetation cover, where most measurements will be returned from plant canopies, some measurements will be returned from the underlying ground surface, resulting in a highly accurate map of canopy height.

Key differences among lidar sensors are related to the laser's wavelength, power, pulse duration and repetition rate, beam size and divergence angle, the specifics of the scanning mechanism, and the information recorded for each reflected pulse. Lasers for terrestrial applications (topography and forest) generally have wavelengths in the range of 900 – 1,064 nanometers (nm), where vegetation reflectance is high. One disadvantage of working in this range of wavelengths is absorption by clouds, which impedes the use of lidar devices during overcast conditions. Bathymetric lidar systems (used to measure elevations under shallow water bodies) make use of wavelengths near 532 nm for better penetration of water.

The power of the laser and size of the receiver aperture determine the maximum flying height, which limits the width of the swath that can be collected in one pass (Wehr and Lohr 1999). The intensity or power of the return signal depends on several factors: the total power of the transmitted pulse, the fraction of the laser pulse that is intercepted by a surface, the reflectance of the intercepted surface at the laser's wavelength, and the fraction of reflected illumination that travels in the direction of the sensor.

The laser pulse returned after intercepting a vegetation canopy will be a complex combination of energy returned from surfaces at numerous distances, the distant surfaces represented later in the reflected signal. The type of information collected from this return signal distinguishes two broad categories of sensors. *Discrete-return or small-footprint* lidar devices measure either one (single return systems) or a small number (multiple return systems) of heights by identifying, in the return signal, major peaks that represent discrete objects in the path of the laser illumination. The distance corresponding to the time elapsed before the leading edge of the peak(s), and sometimes the power of each peak, are typical values recorded by this type of system (Wehr and Lohr 1999). *Waveform-recording or large-footprint* devices record the time-varying intensity of the returned energy from each laser pulse, providing a record of the height distribution of the surfaces illuminated by the laser pulse (Harding et al. 1994, 2001, Dubayah et al. 2000). The small-footprint systems identify, while receiving the return signal, the retention times and heights of major peaks; the large-footprint systems capture the entire signal trace for later processing. Conceptual differences between the two major categories of lidar sensors are illustrated in Figure 1.

Both small-footprint and large-footprint sensors are typically used in combination with instruments for locating the source of the return signal in three dimensions. These include Global Positioning System (GPS) receivers to obtain the position of the platform, Inertial Navigation Systems (INS) to measure the attitude (roll, pitch, and yaw) of the lidar sensor, and angle encoders for the orientation of the scanning mirror(s). Combining this information with accurate time referencing of each source of data yields the absolute position of the reflecting surface, or surfaces, for each laser pulse. One common misconception about lidar data is that they are raster data sets. This is untrue; the data as delivered are typically no more than an ASCII mass point file for each pulse-return combination containing the X, Y, and Z coordinates in the user-specified coordinate system, combining all returns from a particular area results in a point cloud.

There are advantages to both small-footprint and large-footprint lidar sensors. For example, small-footprint systems feature high spatial resolution, made possible by the small diameter of their footprint and the high repetition rates of these systems (as high as 100,000 points per second), which together can yield dense distributions of sampled points. Thus, small-footprint systems are preferred for detailed mapping of ground and canopy surface topography. An additional advantage made possible by this high spatial resolution is the ability to aggregate the data over areas and scales specified during data analysis, so that specific locations on the ground, such as a particular forest inventory plot or even a single tree crown, can be characterized. Finally, small-footprint systems are readily and widely available, with ongoing and rapid development, especially for surveying and photogrammetric applications (Flood and Gutelis 1997). The primary users of these systems are surveyors serving public and private clients, and natural resource managers seeking a cheaper source of high-resolution topographic maps and Digital Elevation Models (DEMs). Most current small-footprint instruments enable an absolute accuracy of the elevation data of 15-20 cm or less, with horizontal position being on the order of 10s of centimeters. A potential drawback is that proprietary data-processing algorithms and established sensor configurations designed for commercial use may not coincide with scientific objectives (Wehr and Lohr 1999).

The advantages of large-footprint lidar include an enhanced ability to characterize canopy structure, the ability to concisely describe canopy information over increasingly large areas, and the availability of global data sets. Examples of large-footprint laser altimeters include MKII (Aldred and Bonnor 1985) and a similar system described in Nilsson (1996), as well as a series of airborne devices developed at NASA's Goddard Space Flight Center, starting with a profiling sensor described by Bufton and colleagues (1991) and including SLICER (Scanning Lidar Imager of Canopies by Echo Recovery; Blair et al. 1994, Harding et al. 1994, 2001), LVIS (Laser Vegetation Imaging Sensor; Blair et al. 1999), and VCL (Vegetation Canopy Lidar; Dubayah et al. 1997) satellite. One advantage of these large-footprint lidar systems is that they record the entire time-varying power of the return signal from all illuminated surfaces and are therefore capable of collecting more information on canopy structure than all but the most spatially dense collections of small-footprint lidar. Additionally, large-footprint lidar integrates canopy structure information over a relatively large footprint and is capable of storing that information efficiently, from the perspective of both data storage and data analysis.

The correspondence between data from each sensor is illustrated in Figure 2, using data collected with a small-footprint lidar at the Upper San Pedro River Basin, Arizona, USA. Section a (up) illustrates the three-dimensional distribution of small-footprint (first-return) data from within a 22 m × 26 m footprint centered on a cottonwood tree approximately 30 m tall. Section b (down) illustrates the distribution of these points as a function of height. Blair and Hofton (1999) and Farid et al. (2008) demonstrated that this vertical distribution of the small-footprint data is closely related to the large-footprint recorded by waveform-recording devices when certain conditions are met — the most important being a high density of samples collected using a very small footprint (on the order of 10-15 cm) so that elevation data can be collected from within very small gaps in the canopy structure. To completely simulate a large-footprint lidar waveform, the vertical distribution of the small-footprint lidar data would have to be corrected for the spatial and temporal distribution of energy within the lidar pulse and receiver response, as described in Blair and Hofton (1999).

APPLICATIONS OF LIDAR REMOTE SENSING

Only a few areas of application for lidar remote sensing have been rigorously evaluated. Numerous other applications are generally considered feasible, but they have not yet been explored; developments in lidar remote sensing are occurring so rapidly that it is difficult to predict which applications are dominant. Currently, applications of lidar remote sensing in ecology and forestry fall into three general categories: (1) Remote sensing of the ground topography; (2) Measurement of the three-dimensional structure and function of vegetation canopies; and (3) Prediction of forest stand structure attributes (such as Leaf Area Index (LAI) and timber volume).

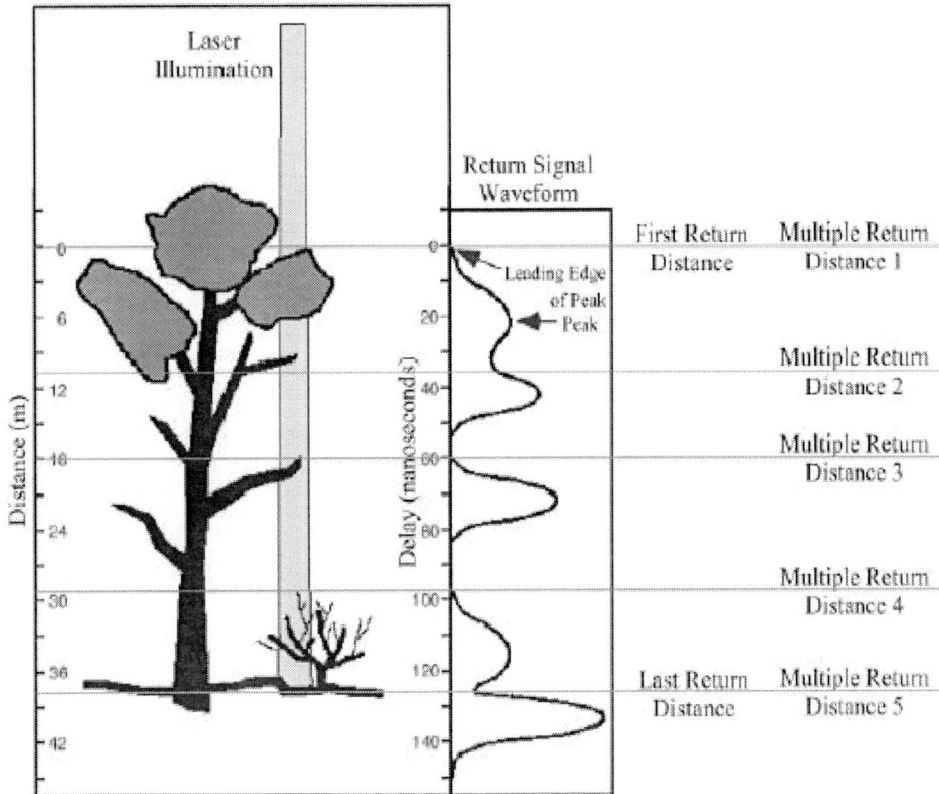

Figure 1. Illustration of the conceptual differences between large-footprint waveform recording and small-footprint discrete-return lidar devices. At the left is the intersection of the laser illumination area, or footprint, with a portion of a tree crown. In the center of the figure is a hypothetical return signal (the large-footprint) that would be collected by a waveform-recording sensor over the same area. To the right of the waveform, the heights recorded by three varieties of small-footprint discrete-return lidar sensors are indicated. First-return lidar devices record only the position of the first object in the path of the laser illumination, whereas last-return lidar devices record the height of the last object in the path of illumination and are especially useful for topographic mapping. Multiple-return lidar, records the height of a small number (five or fewer) of objects in the path of illumination. Figure adapted from Lefsky et al. (2002).

Measuring Vegetation Canopy Structure and Function

In general, the single most important step in lidar mapping of topography involves the deletion of data points returned from vegetation and, in urban areas, buildings.

a)

b)

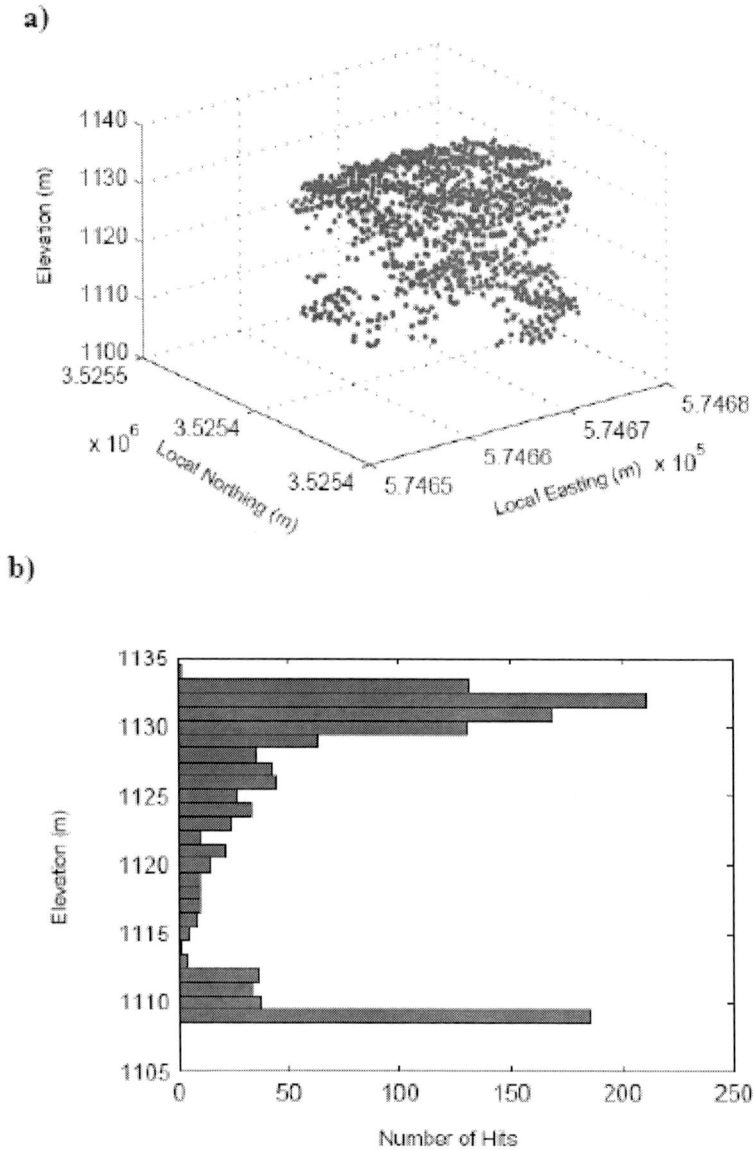

Figure 2. Illustration of the potential for creating synthetic lidar waveforms from small-footprint lidar data. Section a shows the three-dimensional distribution of small-footprint lidar data from within a 22 m × 26 m footprint. Section b shows the vertical distribution of these returns. Figure adapted from Farid et al. (2008), with permission from Elsevier Science.

However, for most forest applications, it is the returns from the vegetation canopy that will be of primary interest. Canopy structure contains a substantial amount of information about the state of development of plant communities (Lefsky et al. 1999a, 1999b) and therefore about canopy function (Hollinger 1989, Brown and Parker 1994) and vegetation-related habitat conditions for wildlife (Hansen and Rotella 2000).

The simplest canopy structure measurements are of canopy height and cover. Canopy heights have been compared with varying accuracy and strength of correlation, to maximum and mean tree height in temperate (Maclean and Krabill 1986), tropical (Nelson et al. 1997,

Drake et al. 2002), boreal (Naesset 1997a, Magnussen and Boudewyn 1998, Magnussen et al. 1999), and riparian (Farid et al, 2006b) forests. In addition, Ritchie and colleagues (1995) found excellent agreement between lidar measurements of height in both temperate deciduous forests and desert scrub.

The latter finding is particularly important, as it indicates that vegetation height measurements can be made accurately even on vegetation of short stature, at least in low-slope environments. There are two general problems in determining vegetation height using lidar data. Determining the exact elevation of the ground surface poses difficulties for both small-footprint and large-footprint waveform-recording lidar. In complex canopies, elevations returned from what appears to be the ground level in fact may be from the understory, if the understory is dense enough to substantially occlude the ground surface. In addition, each type of lidar system presents difficulties in detecting the uppermost portion of the plant canopy. With small-footprint lidar, very high footprint densities are required to ensure that the highest portion of individual tree crowns is sampled. With large-footprint waveform sampling devices, a large footprint is illuminated, increasing the probability that treetops will be illuminated by the laser. However, the top portion of the crown may not be of sufficient area to register as a significant return signal and therefore may not be detected. In either case, the height of the canopy may be underestimated. Estimates of canopy cover have been made using both small-footprint (Farid et al, 2006b) and large-footprint waveform-recording lidar sensors. These estimates are made using the fraction of the lidar measurements that are considered to have been returned from the ground surface (Nelson et al. 1984, Ritchie et al. 1992, 1995, 1996, Weltz et al. 1994, Lefsky 1997), where the measurements are the number of small-footprint returns, or the integrated power of a waveform. In some cases, a scaling factor is needed to correct for the relative reflectance of ground and canopy surfaces at the wavelength of the laser (Lefsky 1997). As with the measurement of canopy height, the definition of the ground surface is a critical aspect of cover determination. If the number of the measurements assigned to the ground return is overestimated cover will be underestimated, and vice versa. Although the height and cover of the canopy surface are useful canopy structure descriptions, there are more detailed measurements that can better describe canopy function and structure. The height distribution of outer canopy surfaces, which quantifies such important features as light gaps (Watt 1947, Canham et al. 1990, Spies et al. 1990), has been manually mapped in several studies (Leonard and Federer 1973, Ford 1976, Miller and Lin 1985). These maps were laboriously made, using devices such as plumb bobs and telescoping rods; with lidar, the process is greatly accelerated (Nelson et al. 1984, Lefsky et al. 1999b). The vertical distribution of all material within the canopy may be inferred, using the foliage-height profile technique (MacArthur and Horn 1969, Aber 1979) adapted for use with large-footprint waveform-recording lidar as the canopy height profile (Lefsky 1997, Harding et al. 2001). Calculation of these height profiles relies on assumptions about the rate of occlusion of canopy surfaces that are not applicable to all forests; however, they have been shown to yield a good approximation in closed-canopy, temperate deciduous forests (Aber 1979, Fukushima et al. 1998, Harding et al. 2001). Lidar data have been used to predict the fractional transmittance of light as a function of height, based on a series of assumptions relating the penetration of the laser light into the canopy to the penetration of natural light into the canopy. Although both the wavelength and orientation of typical laser illumination differ from that of natural illumination, a study (Parker et al. 2001) indicates that lidar can accurately estimate the rate of photosynthetically active radiation absorption and define the location and depth of the zone where the maximum rate of absorption occurs (Parker 1997). Lidar has also been used to predict the aerodynamic properties of plant canopies and landscapes. In modeling airflow over a forest canopy, the aerodynamic roughness length is the height at which the wind speed becomes zero. Menenti and Ritchie (1994) used a profiling laser altimeter to predict aerodynamic roughness length of complex landscapes containing a mixture of grassland, shrub, and woodland areas, and found good agreement with field estimates. The techniques described so far use lidar data to make measurements of canopy structure that had been made with technologically simpler and more time-consuming methods. Lidar ability to rapidly measure the three-dimensional structure of canopies should stimulate the development of new systems of canopy description. One such system, the canopy volume method (CVM), is the first to take advantage of the ability of a waveform-recording sensor (SLICER) to directly measure the three-dimensional distribution of canopy structure. This approach led to a better understanding of the structure of the old-growth forest canopy, new visualizations of the multiple canopy aspect of old-growth development, and improved estimates of forest stand structure.

Predictions of Forest Stand Structure

Lidar data have been used to predict biophysical characteristics of forest communities (Dubayah and Drake 2000). Although the following studies may not by themselves constitute forest research, they lay the groundwork for future studies that use these relationships to map biophysical variables over large extents, making possible a new class of large-scale forest research. Prediction of forest stand structure using small-footprint lidar had its start in the work of Maclean and Krabill (1986), who adapted a photogrammetric technique — the canopy profile cross-sectional area — to the interpretation of lidar data. Nelson et al. (1988) successfully predicted the volume and biomass of southern pine (Pinus taeda, P. elliotti, P. echinata, and P. palustris) forests using several estimates of canopy height and cover from small-footprint lidar, explaining between 53% and 65% of variance in field measurements of these variables. Later work by Nelson et al. (1997) in tropical wet forests at the La Selva Biological Station obtained similar results for prediction of basal area, volume, and biomass. They also developed a canopy structure model that led to greater understanding of the optimal spatial configuration of field sampling for comparison with profiling lidar data. Naesset (1997b) explained 45%–89% of variance in stand volume in stands of Norway spruce (Picea abies) and Scots pine (Pinus sylvestris), using measurements of maximum and mean canopy height and cover. Nilsson (1996) adapted a bathymetric lidar system for use in forest inventory, and successfully predicted timber volume for stands of even-aged Scots pine (P. sylvestris). He used the height and the total power of each waveform as independent variables, and explained 78% of variance. Lefsky and colleagues (1999a) used data from SLICER to predict aboveground biomass and basal area in eastern deciduous forests using indices derived from the canopy height profile. Of particular note, they found that relationships between height indices and forest structure attributes (basal area and aboveground biomass) could be generated using field estimates of the canopy height profiles, and applied directly to the lidar-estimated profiles, resulting in unbiased estimates of forest structure. Means and colleagues (1999) applied similar methods to evaluate 26 plots in forests of Douglas-fir and western hemlock at the H. J.Andrews Experimental Forest. They found that very accurate estimates of basal area, aboveground biomass, and foliage biomass could be made using lidar height and cover estimates. Farid et al (2006a, 2006b, and 2008) used multi-return small-footprint lidar to identify riparian tree species, age, and canopy characteristics. Stepwise multiple regressions (canopy height, height of median energy, ground return ratio, and canopy return ratio) were performed to predict ground-based measures of stand structure from riparian canopy structure indices (LAI). The method used four metrics (canopy height, height of median energy, ground return ratio, and canopy return ratio) that were derived by synthetically construction of a large footprint lidar waveform from the airborne small-footprint lidar data (see Figure 3). Farid et al (2008) concentrated on individual cottonwood trees to develop the relationships to estimate LAI for riparian water use estimates and may not be applicable to dense, overlapping canopies. Additionally, lidar cannot provide information on stomatal control with also regulates riparian cottonwood water use so independent estimates or typical ranges of canopy level stomatal resistance will be required. However, strategically acquired lidar data and derived spatially explicit LAI measurements, offer significant potential to improve riparian water use estimates. Future research will investigate how well lidar can derive LAI in more complex and interacting canopies.

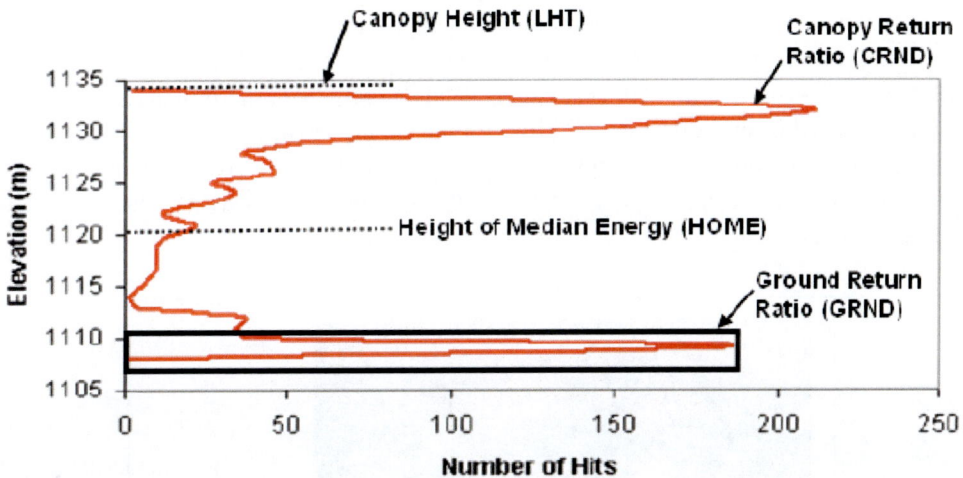

Figure 3. Metrics derived from synthetic large footprint lidar waveforms. These metrics were then used to estimate LAI for different age classes of cottonwoods. Figure adapted from Farid et al. (2008), with permission from Elsevier Science.

CONCLUSIONS

Lidar remote sensing recently has become available as a research tool, and it has yet to become widely available. Nevertheless, it has already been shown to be an extremely accurate tool for measuring topography, vegetation height, and cover, as well as more complex attributes of canopy structure and function. Additionally, the basic canopy structure measurements made with lidar sensors have been shown to provide highly accurate estimates of important forest stand structure indices, such as leaf area index and aboveground biomass. Because the basic measurements made by lidar sensors are directly related to vegetation structure and function, we expect that these findings will continue to be corroborated in a variety of biomes, with similar results. The availability of lidar data will increase with the broader use of airborne sensors for topographic and forest canopy mapping. As data availability grows, a variety of applications will become feasible. It is likely that lidar will be useful in detecting habitat features associated with particular species, including those that are rare or endangered. For instance, the large open-grown trees and associated old-growth habitat that serve as nesting habitat for marbled murrelets (Hamer and Nelson 1995) should be readily identifiable from lidar data. Another likely application of lidar data is the identification of forest areas with accumulations of fuels that make them particularly susceptible to large, especially damaging fires (Agee 1993). Lidar's ability to discriminate the spatial pattern as well as the total volume of materials within a forest canopy would be especially useful for identifying, at the least, classes of forest structure that are associated with varying fire behavior. For instance, lidar should enable the detection of "ladder" fuels, which provide a pathway for ground-level fires to reach the upper canopy and cause more damaging crown fires. Additionally, the ability to identify the size and depth of canopy gaps should allow estimation of the quantity of large woody fuels associated with the creation of those gaps. Lidar remote sensing shows great potential for integration with forest research. It directly measures the physical attributes of vegetation canopy structure, which are highly correlated with the basic plant community measurements of interest to ecologists. The detailed measurement and modeling of canopies has largely been the province of specialists. By reducing the time and effort associated with measuring canopy structure, lidar can foster the wider incorporation of a canopy science perspective into forest research and put vegetation canopy structure squarely at the center of efforts to measure and model global carbon dynamics.

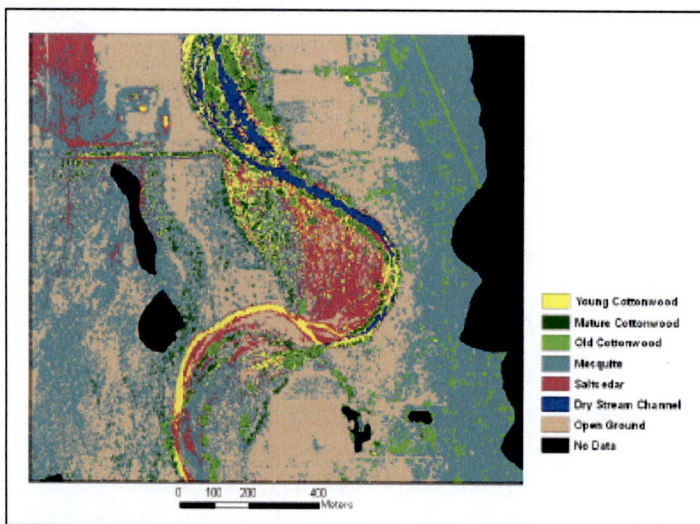

Figure 4. Classified lidar image, showing three cottonwood age classes, mesquite, saltcedar, dry stream channel, and open ground. Adapted from Farid et al (2006a), with permission from Canadian Journal of Remote Sensing.

REFERENCES

Aber JD. 1979. Foliage-height profiles and succession in northern hardwood forests. *Ecology* 60: 18–23.

Agee JK. 1993. Fire Ecology of Pacific Northwest Forests. Washington (DC): Island Press.

Aldred A, Bonnor G. 1985. Application of airborne lasers to forest surveys. Chalk River (Ontario, Canada): Petawawa National Forestry Institute, *Canadian Forestry Service.* Information Report PI-X-51.

Bachman CG. 1979. Laser Radar Systems and Techniques. Norwood (MA): Artech House.

Blair JB, Coyle DB, Bufton JL, Harding DJ. 1994. Optimization of an airborne laser altimeter for remote sensing of vegetation and tree canopies. Pages 939–941 in Proceedings of the International Geosciences Remote Sensing Symposium. Pasadena (CA): *California Institute of Technology.*

Blair JB, Hofton MA. 1999. Modeling laser altimeter return waveforms over complex vegetation using high-resolution elevation data. *Geophysical Research Letters* 26: 2509–2512.

Brown MJ, Parker GG. 1994. Canopy light transmittance in a chronosequence of mixed-species deciduous forests. *Canadian Journal of Forestry Research* 24: 1694–1703.

Bufton JL, Garvin JB, Cavanaugh JF, Ramos-Izquierdo L, Clem TD, Krabill WB. 1991. Airborne lidar for profiling of surface topography. *Optical Engineering* 30: 72–78.

Canham CD, Denslow JS, Platt WJ, Runkle JR, Spies TA, White PS. 1990. Light regimes beneath closed canopies and tree-fall gaps in temperate and tropical forests. Canadian *Journal of Forestry Research* 20: 620–631.

Drake, J.B., Dubayah, R.O., Clark, D.B., Knox, R.G., Blair, J.B., Hofton, M.A., Chazdon, R.L., Weishampel, J.F., Prince, S.D., 2002. Estimation of tropical forest structural characteristics using large-footprint lidar. *Remote Sensing of Environment* 79, 305-319.

Dubayah R, Blair JB, Bufton JL, Clark DB, JaJa J, Knox R, Luthcke SB, Prince S, Weishampel J. 1997. The vegetation canopy lidar mission. Pages 100-112 in Proceedings of Land Satellite Information in the Next Decade, II: Sources and Applications. Bethesda (MD): American Society of Photogrammetry and Remote Sensing.

Dubayah R, Knox R, Hofton M, Blair JB, Drake J. 2000. Land surface characterization using lidar remote sensing. Pages 25–38 in Hill MJ, Aspinall RJ, eds. Spatial Information for Land Use Management. Singapore: *International Publishers Direct.*

Dubayah RO, Drake JB. 2000. Lidar remote sensing for forestry. *Journal of Forestry* 98: 44–46.

Farid, A., Rautenkranz, D., Goodrich, D.C., Marsh, S.E., Sorooshian, S., 2006a. Riparian vegetation classification from airborne laser scanning data with an emphasis on cottonwood trees. *Canadian Journal of Remote Sensing* 32, 15-18.

Farid, A., Goodrich, D.C., Sorooshian, S., 2006b. Using airborne lidar to discern age classes of cottonwood trees in a riparian area. Western Journal of Applied Forestry 21, 149-158.

Farid, A., Goodrich, D.C., Bryant, R., Sorooshian, S., 2008. Using airborne lidar to predict leaf area index in cottonwood trees and refine riparian water-use estimates. *Journal of Arid Environments*, Vol. 72, No. 1, pp. 1-15.

Flood M, Gutelis B. 1997. Commercial implications of topographic terrain mapping using scanning airborne laser radar. *Photogrammetric Engineering and Remote Sensing* 63: 327–366.

Ford ED. 1976. The canopy of a Scots pine forest: Description of a surface of complex roughness. *Agricultural and Forest Meteorology* 17: 9–32.

Fukushima Y, Hiura T, Tanabe S. 1998. Accuracy of the MacArthur-Horn method of estimating a foliage profile. *Journal of Agricultural Forest Meteorology* 92: 203–210.

Hamer TE, Nelson SK. 1995. Characteristics of marbled murrelet nest trees and nesting stands.Pages 69–82 in Ralph CJ, Hunt GL, Raphael MG, Platt JF, eds. Ecology and Conservation of the Marbeled Murrelet. Albany (CA): US Department of Agriculture, Forest Service, Pacific Southwest Research Station. *General Technical Report* PSW-152.

Hansen A, Rotella JJ. 2000. Bird responses to forest fragmentation. Pages 201–219 in Knight RL, Smith FW, Romme WH, Buskirk SW, eds. Forest Fragmentation in the Southern Rockies. Boulder: University Press of Colorado.

Harding DJ, Blair JB, Garvin JB, Lawrence WT. 1994. Laser altimetry waveform measurement of vegetation canopy structure. Pages 1251–1253 in Proceedings of the International Remote Sensing Symposium 1994. Pasadena (CA): California Institute of Technology.

Harding DJ, Lefsky MA, Parker GG, Blair JB. 2001. Lidar altimeter measurements of canopy structure: Methods and validation for closed canopy, broadleaf forests. *Remote Sensing of Environment* 76: 283–297.

Hollinger DY. 1989. Canopy organization and photosynthetic capacity in a broad-leaved evergreen montane forest. *Functional Ecology* 3: 53–62.

Lefsky MA, Cohen WB, Geoffrey GP, Harding DJ 2002. Lidar Remote Sensing for Ecosystem Studies. *BioScience,* Vol. 52, No. 1, pp. 19-30.

Lefsky MA, Cohen WB, Acker SA, Spies TA, Parker GG, Harding D. 1999b. Lidar remote sensing of biophysical properties and canopy structure of forest of Douglas-fir and western hemlock. *Remote Sensing of Environment* 70: 339–361.

Lefsky MA, Harding D, Cohen WB, Parker GG. 1999a. Surface lidar remote sensing of the basal area and biomass in deciduous forests of eastern Maryland, USA. *Remote Sensing of Environment* 67: 83–98.

Lefsky MA. 1997. Application of Lidar Remote Sensing to the Estimation of Forest Canopy and Stand Structure. PhD dissertation. University of Virginia, Charlottesville,VA.

Leonard RE, Federer CA. 1973. Estimated and measured roughness parameters for a pine forest. *Journal of Applied Meteorology* 12: 302–307.

MacArthur RH, Horn HS. 1969. Foliage profile by vertical measurements. *Ecology* 50: 802–804.

Maclean GA, Krabill WB. 1986. Gross-merchantable timber volume estimation using an airborne LIDAR system. *Canadian Journal of Remote Sensing* 12: 7–18.

Magnussen S, Boudewyn P. 1998. Derivations of stand heights from airborne laser scanner data with canopy-based quantile estimators. *Canadian Journal of Forestry Research* 28: 1016–1031.

Magnussen S, Eggermont P, LaRiccia VN. 1999. Recovering tree heights from airborne laser scanner data. *Forest Science* 45: 407–422.

Means JE, Acker SA, Harding DJ, Blair JB, Lefsky MA, Cohen WB, Harmon M, McKee WA. 1999. Use of large-footprint scanning airborne lidar to estimate forest stand characteristics in the western Cascades of Oregon. *Remote Sensing of Environment* 67: 298–308.

Menenti M, Ritchie JC. 1994. Estimation of effective aerodynamic roughness of Walnut Gulch Watershed with laser altimeter measurements. *Water Resources Research* 30: 1329–1337.

Miller DR, Lin JD. 1985. Canopy architecture of a red maple edge stand measured by the point drop method. Pages 59–70 in Hutchinson BA, Hicks BB, eds. The Forest–Atmosphere Interaction. Dordrecht (Netherlands): Reidel Publishing.

Naesset E. 1997a. Determination of mean tree height of forest stands using airborne laser scanner data. *ISPRS Journal of Photogrammetry and Remote Sensing* 52: 49–56.

Naesset E. 1997b. Estimating timber volume of forest stands using airborne laser scanner data. *Remote Sensing of Environment* 61: 246–253.

Nelson R, Oderwald R, Gregoire TG. 1997. Separating the ground and airborne laser sampling phases to estimate tropical forest basal area, volume, and biomass. *Remote Sensing of Environment* 60: 311–326.

Nelson RF, Krabill WB, Tonelli J. 1988. Estimating forest biomass and volume using airborne laser data. *Remote Sensing of Environment* 24: 247–267.

Nelson RF, Krabill WB, Maclean GA. 1984. Determining forest canopy characteristics using airborne laser data. *Remote Sensing of Environment* 15: 201–212.

Nilsson M. 1996. Estimation of tree heights and stand volume using an airborne lidar system. *Remote Sensing of Environment* 56: 1–7.

Parker GG, Lefsky MA, Harding DJ. 2001. PAR transmittance in forest canopies determined from airborne lidar altimetry and from in-canopy quantum measurements. *Remote Sensing of Environment* 76: 298–309.

Parker GG. 1997. Canopy structure and light environment of an old-growth Douglas fir/western hemlock forest. *Northwest Science* 71: 261–270.

Ritchie JC, Everitt JH, Escobar DE, Jackson TJ, Davis MR. 1992. Airborne laser measurements of rangeland canopy cover and distribution. *Journal of Range Management* 45: 189–193.

Ritchie JC, Humes KS, Weltz MA. 1995. Laser altimeter measurements at Walnut Gulch watershed, Arizona. *Journal of Soil and Water Conservation* 50: 440–442.

Ritchie JC, Menenti M, Weltz MA. 1996. Measurements of land surface features using an airborne laser altimeter: The HAPEX-Sahel experiment. *International Journal of Remote Sensing* 17: 3705–3724.

Spies TA, Franklin JF, Klopsch M. 1990. Canopy gaps in Douglas-fir forests of the Cascade Mountains. *Canadian Journal of Forestry Research* 5: 649–658.

Watt AS. 1947. Pattern and process in the plant community. Journal of Ecology 35: 1–22.

Wehr A, Lohr U. 1999. Airborne laser scanning—an introduction and overview. ISPRS *Journal of Photogrammetry and Remote Sensing* 54: 68–82.

Weltz MA, Ritchie JC, Fox HD. 1994. Comparison of laser and field-measurements of vegetation height and canopy cover. *Water Resources Research* 30: 1311–1319.

In: Forest Canopies: Forest Production, Ecosystem…   ISBN 978-1-60741-457-5
Editor: J. D. Creighton and P. J. Roney   © 2009 Nova Science Publishers, Inc.

Chapter 5

SOIL ORGANIC CARBON DYNAMICS OF DIFFERENT LAND USE IN SOUTHEAST ASIA

Minaco Adachi[1] and Hiroshi Koizumi[2]

[1] National Institute for Agro-Environmental Science, Dept. of Global Resources,
3-1-3 Kannondai, Tsukuba 305-8604 Japan
[2] River Basin Research Center, Gifu University, Yanagido 1-1,
Gifu 501-1193, Japan

ABSTRACT

Soil respiration, CO_2 efflux from the soil surface, is an important process of the carbon (C) cycle in terrestrial ecosystems. Soil respiration includes many processes involving biotic factors, such as respiration from roots and microorganisms, along with abiotic factors and various temporal and spatial factors. Many researchers have examined soil respiration of various ecosystems. Recently, tropical forests have been converted into secondary forests or agricultural forests. This land use change might strongly affect the global C cycle; nevertheless, few data are available to reflect land use effects on dynamics of soil organic carbon (SOC) in Southeast Asia. Therefore, we established study sites at four different ecosystems (primary forest, secondary forest, oil palm plantation, and rubber plantation) in the Pasoh area of Malaysia in Southeast Asia. This study was designed to determine spatial and temporal variations (diurnal and seasonal change) of the soil respiration rate and to estimate the annual C efflux from soil and dynamics of SOC in different ecosystems.

Seasonal data suggest that the soil respiration rate is negatively correlated with soil water contents in the primary forest, secondary forest, and rubber plantation. The soil water content shows a negative correlation with the gaseous phase content. The gaseous phase content shows a positive correlation with soil respiration rate at all sites.

---

[1] Corresponding author: Minaco Adachi, Center for Global Environmental Center, National Intitute for Environmental Studies, 16-2 Onogawa, Tsukuba, Ibaraki 305-8506 Tel: +81-29-850-2567, Fax: +81-29-850-2960, E-mail: adachi.minaco@nies.go.jp.

The annual C efflux from soil was estimated as 16.9–19.2 t C ha^{-1} in the primary forest, 17.5–18.5 t C ha^{-1} in the secondary forest, 14.3–14.5 t C ha^{-1} in the oil palm plantation, and 9.0–11.2 t C ha^{-1} in the rubber plantation. Moreover, we estimated the annual SOC budgets using the three-box model. Results suggest that the biomass of dead roots, turnover time, and contribution of heterotrophic respiration are important factors for accurate evaluation of soil C dynamics and budgets.

Keywords: soil respiration rate, soil water content, tropical forest, oil palm plantation, rubber plantation, box-model

INTRODUCTION

The amount of CO_2 exchange between the atmosphere and terrestrial ecosystems has been argued in many studies. A recent study has estimated that the annual net primary production of tropical regions accounts for 32% of global terrestrial photosynthesis (Field *et al.*, 1998). Tropical forests are expected to be a large carbon sink because of their high productivity. Recent studies of carbon cycles in tropical rain forests are exemplified by those of Malhi *et al.* (1999) and Malhi and Grace (2000). Malhi and Grace (2000) calculated that the tropical forest in Manaus, Brazilian Amazon is a sink of 5.9 t C ha^{-1}. They also concluded that the net carbon balance differed according to the estimation method or tropical region, and pointed out that net biotic carbon sinks can be overestimated or underestimated because of insufficient sample areas for statistical analyses in Asia and Africa. Tropical forests in Asia are rapidly being converted into secondary forests or plantations; deforestation to create permanent croplands has accounted for approximately 75% of the total CO_2 emissions from tropical Asia in the 1980s (Houghton and Hackler, 1999). Annual C flux to the atmosphere from changes in land use in tropical Asia was 0.88 Pg C y^{-1} during the 1980s and 1.09 Pg C y^{-1} during the 1990s (Houghton, 2003).

Soil has an important role in carbon storage in terrestrial ecosystems. Soil carbon pools contain 2500 Gt carbon, which includes about 1550 Gt of soil organic carbon (SOC) and 950 Gt of soil inorganic carbon (Lal, 2004). In fact, SOC sequestration depends on vegetation and the distribution of root biomass (e.g. Jobbagy and Jackson, 2000), tillage (e.g. Balesdent *et al.*, 2000), and soil structure (e.g. Golchin *et al.*, 1994). The most SOC in the top 3 m depth of soil is found in tropical evergreen forests and tropical grasslands and savannas, their amounts have been evaluated respectively as 158 and 146 Pg C (Jobbagy and Jackson, 2000). The litter supply and decomposition rates should be determined to clarify SOC sequestration. Tropical forests contain large amounts of C within their vegetation and soil, equivalent to 37% of global terrestrial C pools (Dixon *et al.*, 1994). The CO_2 efflux from soil (soil respiration) is an important component of the C balance in terrestrial ecosystems. The soil respiration rate, which includes root respiration and heterotrophic respiration, is affected by various soil environmental factors: soil temperature, soil water content, biotic activity in soil, vegetation, soil chemistry, and physical characteristics. Soil respiration varies temporally and spatially; soil temperature and water content are key factors which are responsible for variation in soil respiration. Therefore, many studies have measured soil respiration in many of the world's ecosystems (Raich and Schlesinger, 1992, Raich and Tufekcioglu, 2000).

Some reports have described the relationship between soil respiration and the underground environment, e.g., root biomass (Fang *et al.*, 1998; Søe and Buchmann, 2005; Adachi *et al.*, 2006) and soil microbial biomass (Neergaard *et al.*, 2002; Adachi *et al.*, 2006). In temperate regions, soil temperature is the most important determinant of temporal variation in soil respiration (Davidson *et al.*, 1998; Xu and Qi, 2001; Søe and Buchmann, 2005; Mo *et al.*, 2005). For tropical regions, variation of soil temperature is less than temperate regions, therefore various factors measured; soil temperature and soil water content (e.g., Davidson *et al.*, 2000; Kiese and Butterbach-Bahl, 2002; Sotta *et al.*, 2004; Hashimoto, 2005; Schwendenmann and Veldkamp, 2006), CO_2 concentration in soil or gas diffusivity (Schwendenmann *et al.*, 2003; Hashimoto and Kamatsu, 2006), trees basal area (Sotta *et al.*, 2004), fine root biomass (Silver et al., 2005; Adachi et al., 2006) and microbial community (Cleveland *et al.*, 2007). On the other hand, Ohashi *et al.* (2007) indicated existence of hot spots of soil respiration. Consequently, many studies of soil respiration have evaluated the carbon flux from soils of different ecosystems:, high Arctic area (e.g., Bekku *et al.*, 2004; Elberling, 2007), boreal forests (e.g., Rayment and Jarvis, 2000; Søe and Buchmann, 2005), temperate forests (e.g., Xu and Qi, 2001; Mo *et al.*, 2005), tropical rain forests (e.g., Schwendenmann et al., 2003; Hashimoto, 2005; Adachi *et al.*, 2005, 2006; Ohashi *et al.*, 2007), agricultural land (e.g., Stoyan *et al.*, 2000; Nishimura *et al.*, 2008), and plantations (e.g., Fang *et al.*, 1998; Epron *et al.*, 2004, 2006). Compartment models of SOC dynamics include many factors (e.g., autotrophic respiration rate, litter fall, decomposition rate, dead root, and litter quality), however, assessing these factors in the field is difficult.

For this study, the objectives were (1) to identify seasonal variation in soil respiration rate, (2) to clarify the relationship between the soil respiration rate and soil environmental factors, and (3) to estimate the annual carbon efflux from soil and SOC dynamics in different ecosystems in tropical Southeast Asia using box models.

METHODS

Site Description

Our study was conducted in primary and secondary forests of the Pasoh Forest Reserve in the state of Negeri Sembilan on the Malaysian Peninsula (2°5′N, 102°18′W, Figure1), and in an oil palm plantation adjacent to the Pasoh Forest Reserve. The mean monthly maximum air temperature is 32.5 ± 1.1°C and the minimum air temperature is 22.5 ± 0.5°C. Mean annual precipitation is 1450–2341 mm y^{-1} in the Pasoh Forest Reserve (Adachi *et al.*, 2006). The primary and secondary forests are dominated by Dipterocarpaceae (Tang *et al.*, 1996).The secondary forest site is located in an area in which all trees with trunk diameter at breast height (DBH) of 45 cm had been logged selectively in 1958, and has since undergone natural regeneration (Okuda *et al.*, 2003). In the oil palm plantation plot, Elaeis guineensis (African oil-palm) seedlings were planted in 1976; the site has been fertilized and weeded annually. We started to measure soil respiration in the oil palm plantation in August 2000, but this plantation's trees were cut down in October 2001; seedlings were planted after one year. For that reason, in June 2002, we established a new plot near the previous plot. In the rubber plantation plot, Hevea brasiliensis seedlings were planted in 1989.

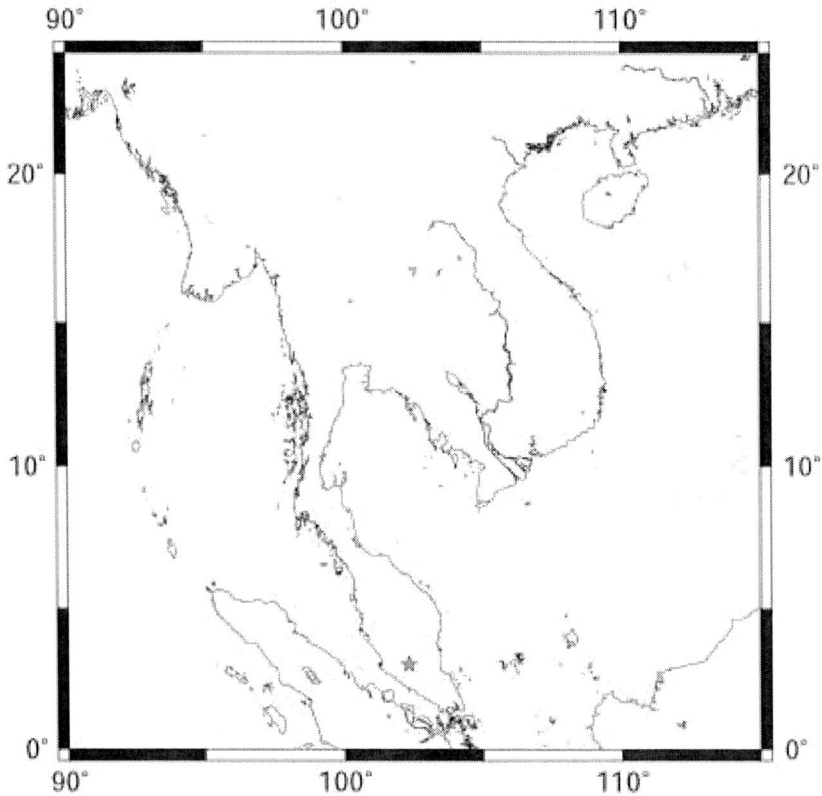

Figure 1. Location of the Pasoh Forest Reserve.

Yamashita et al. (2003) reported details of soil types of the area. The pH of the topsoil (0–5 cm) was 3.8 ± 0.2 (mean $\pm$ S.D, n = 28) in the primary forest, 4.2 ± 0.2 (n = 32) in the secondary forest, 4.7 ± 0.4 (n = 16) in the oil palm plantation (Adachi et al., 2006), and 5.4 ± 0.3 (n = 10) in the rubber plantation (Yashiro et al., 2007).

Soil Respiration Rate

Soil respiration, soil temperature, and soil water content were measured every 3–4 months during May 2000 – December 2003 in the primary and secondary forest, during August 2000 – December 2003 in the oil palm plantation, and during October 2002 – December 2003 in the rubber plantation. The soil respiration rate was measured in a grid pattern at 16 points in a 64 m^2-quadrat (under the canopy) in all plots (n =16). We estimated the required sample sizes for estimating large-scale soil respiration (for areas of 1–2 ha) in four ecosystems. According to our sample size analysis, the number of measurement points necessary to determine the mean soil respiration rate at each site with an error from the mean of no more than 10% was 67–85 at the 95% probability level (Adachi et al., 2005).

To minimize the effects of chamber installation, 24 h before the soil respiration measurements were made, a soil collar (5 cm high and 13 cm diameter) was set into the soil to a depth of about 1 cm at each sampling point, taking care not to disturb the soil structure.

The soil respiration rate was measured using a portable soil respiration rate measuring system (LI-6400; Li-Cor Inc., Lincoln, NB, USA) fitted with a soil respiration chamber (6400-09; Li-Cor Inc., NB, USA). At the same location used for measurement of soil respiration, soil and plant roots were sampled to a depth of 10 cm and 13 cm diameter during 4–7 September 2000 at the three ecosystem sites. All roots were divided from the soil with the hand. Root respiration was measured with a capped LI-6400 soil chamber immediately after roots had been removed from soil at 20 °C. Coarse and fine roots of respiration were measured separately, and all root samples were dried for 48h at 80°C and weighed. Root respirations per dry weight were calculated, and autotrophic respiration was estimated that total soil respiration minus root respiration.

Environment Factors

Simultaneously with the soil respiration measurements, soil temperatures at depths of 1 cm and 5 cm and soil water content at 5 cm depth were measured at each point. Soil temperatures were measured using a thermometer (TM-150; Custom, Tokyo, Japan). The soil water contents were measured using a time-domain reflectometer (TDR, TRIME-FM; IMKO Micromodultechnik GmbH, Ettlingen, Germany).

When we evaluated the annual carbon flux from soil in the four ecosystems, we used soil water content data for every 12 h during September 2000 – August 2001 in the primary and secondary forest and oil palm plantation, and during December 2002 – January 2003 in the rubber plantation.

Soil Physics

After we measured soil respiration on October, 2002, we collected soil at the same place using a 100 ml soil sampler. Volumes of soil solids were measured using a three-phase meter (DIK-1121; Daiki Rika Kogyo Co. Ltd., Japan). Soil samples were dried at 105°C at 48 h. Then they were weighed; we calculated the respective volumes of soil liquid and gas (Hillel, 1998). For the present study, soil water contents were volume wetness, which were measured using TDR; the volume of liquid was the volumetric water contents of soil core samples.

For soil temperature and water contents monitoring, thermometer (Titbit, Onset, U.S.A) and Theta probe with data logger (Type ML2x, Dynamax, Texas, U.S.A) were installed soil in primary and secondary forest and oil palm plantation. Soil temperatures were measured at 1cm and 5cm depth every 1 hour, and soil water contents were measured at 5 cm depth every 12 hours from April 2000 to November 2003 (Figure2).

RESULTS

Figure 3 presents mean soil respiration and soil water contents for the four ecosystems during four years.

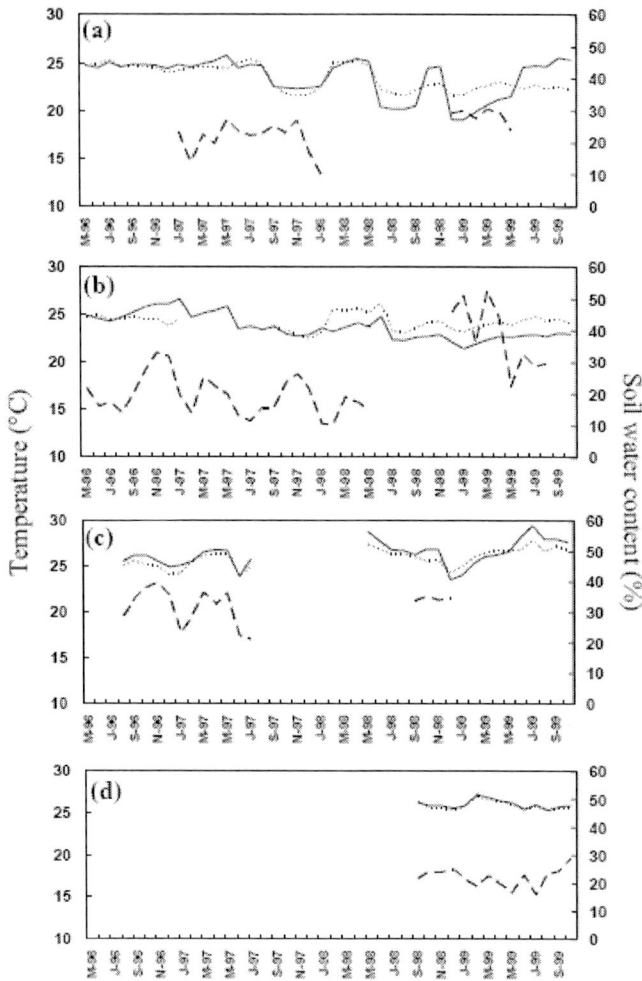

Figure 2. Seasonal change of environmental factors in the study site from May 2000 to September 2003. Solid lines show air temperature, dashed lines show soil temperature (1cm depth), and bold broken lines show soil water content. (a); primary forest, (b); secondary forest, (c); oil palm plantation, (d); rubber plantation.

Mean soil respiration rates were 676–1041 mg CO_2 m^{-2} hr^{-1} in the primary forest, 609–1177 mg CO_2 m^{-2} hr^{-1} in the secondary forest, 108–1019 mg CO_2 m^{-2} hr^{-1} in the oil palm plantation and 256–615 mg CO_2 m^{-2} hr^{-1} in the rubber plantation. Soil water contents also changed seasonally: 13–28% in the primary forest, 10–28% in the secondary forest, 9–39% in the oil palm plantation, and 25–33% in the rubber plantation. The soil respiration rates in the rubber plantation were always lower than in the primary forest and secondary forest, the rates were 35–92% of those of the primary forest. Soil respiration rates and soil water contents varied seasonally, however, soil temperature varied only slightly in all plots. All data of soil respiration rates and soil water contents are presented in Figure 3; the soil respiration rate shows a negative relation to soil water contents in primary and secondary forest and rubber plantation (Figure 4). One possible reason for this result is that gas diffusivity decreased with conditions of high soil water contents.

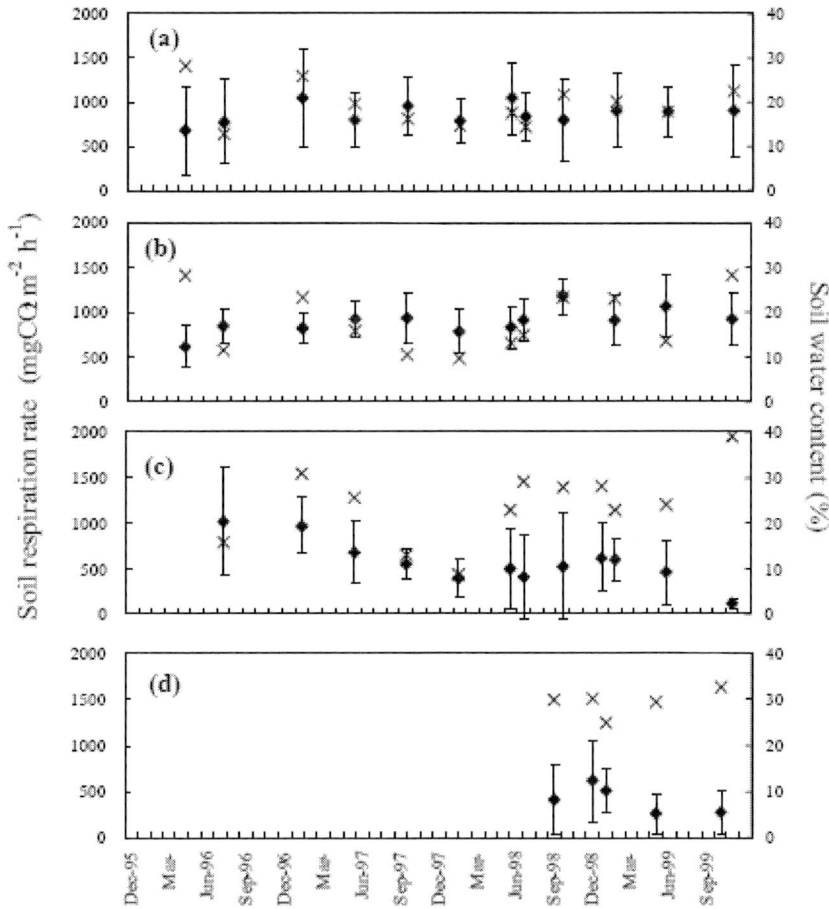

Figure 3. Seasonal variation in soil respiration and soil water content: (a) primary forest, (b) secondary forest, (c) oil palm plantation, (d) rubber plantation.

Figure 5 presented that relationship among soil respiration rate, the volume of soil air, and soil water content, as measured with TDR in four ecosystems. This result suggests that the negative correlation of soil respiration and soil water contents reflects the soil physics, especially the volume of soil air. Correlation coefficients obtained for respective plots are presented in Table 1. Volumes of soil solid phase in plantation plots were significantly higher than in primary and secondary forests. On the other hand, volumes of soil air phase in plantation plots were significantly lower than in forest plots (Table 2).

We estimated the annual carbon efflux from soil using the relationship between soil respiration and soil water contents: the results were 18 t C ha^{-1} yr^{-1} in primary and secondary forests, 14 t C ha^{-1} yr^{-1} in oil palm plantation, and 10 t C ha^{-1} yr^{-1} in rubber plantations (Table 3).

The SOC was estimated using some box models (Figure 6 from Itoh, 2002). SOC sequestration represents the difference between heterotrophic respiration and the carbon supply from litter. The three-box model included dead root biomass to soil as a kind of litter.

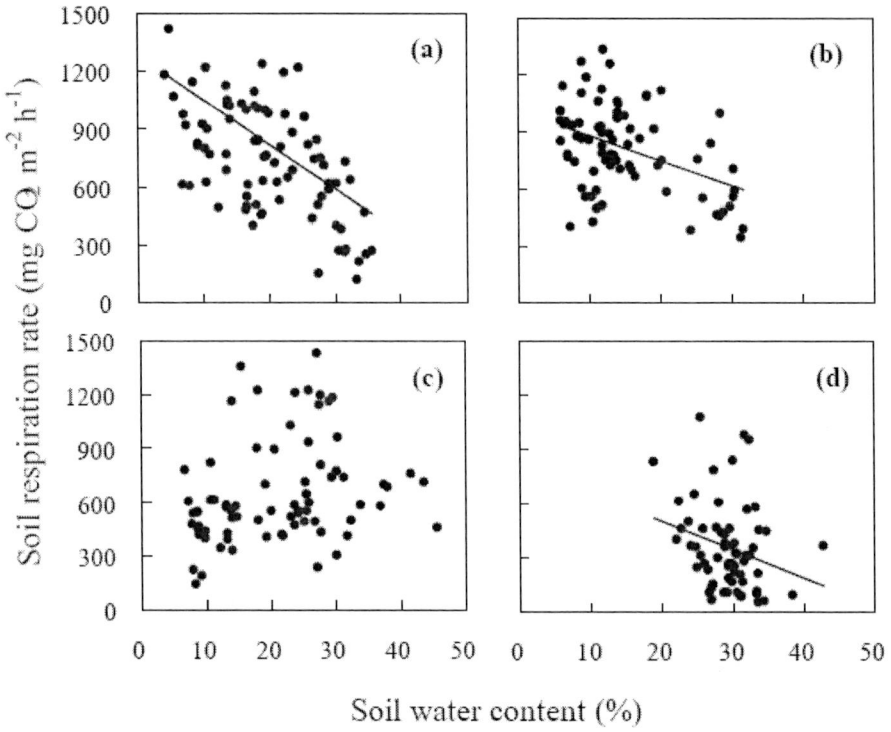

Figure 4. Relationships between soil respiration rate and soil water content by seasonal data: (a) primary forest, (b) secondary forest, (c) oil palm plantation, (d) rubber plantation. The respective regression equations suggest significant relationships: (a) y = -23.192x + 1288.5 R2 = 0.211; (b) y = -13.272x + 1013.9 R2 = 0.1569; (d) y = -15.598x + 812.28 R2 = 0.0662.

The SOC sequestration in primary forest and oil palm plantation were estimated using these box models (Table 4), because we could not have information for calculations in rubber plantation. On the one-box model, primary forest soil lost carbon in -5.3 t C ha^{-1} y^{-1}. However, the primary forest was predicted to have a large underground biomass. Therefore, SOC sequestration was almost zero on the three-box model because the primary forest has a large aboveground biomass.

DISCUSSION

Seasonal Variation in Soil Respiration

Spatial variations in soil water contents were important for estimation of annual carbon efflux from soil at the study site. In the tropical forest, results of some studies support that soil temperature only slightly changes throughout the year. For that reason, seasonal variation in soil respiration depends on rainfall (Kursar, 1989; Davidson et al., 2000). Mean precipitation at the study site shows no distinct difference between rainy and dry seasons. However, soil water contents at 5 cm depth showed a seasonal change. Generally, soil respiration has a strong positive relationship with soil temperature.

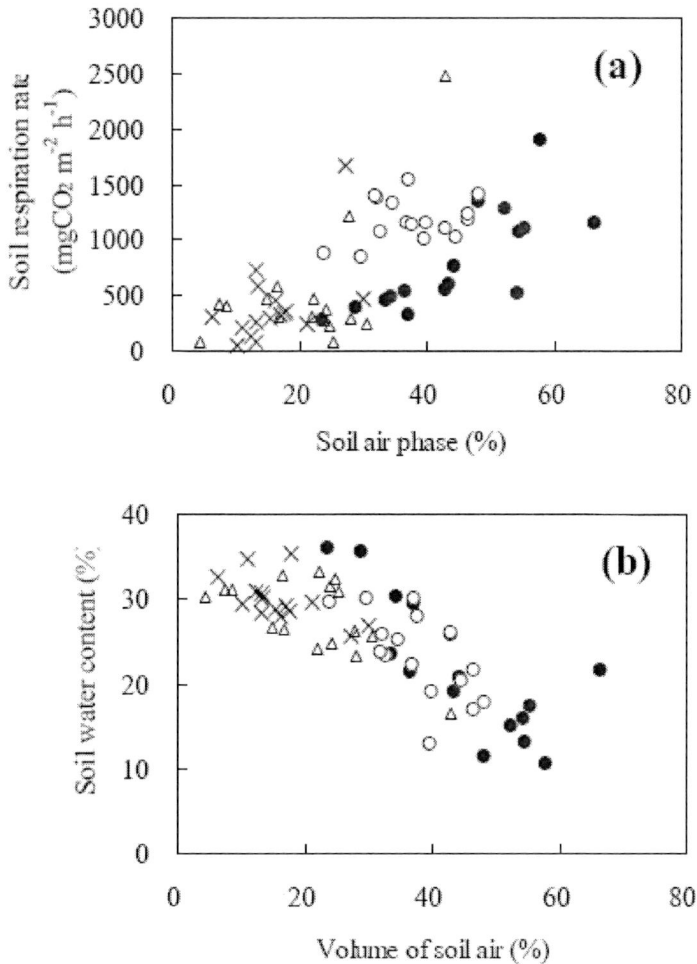

Figure 5. The relationships between (a) soil respiration and volume of soil air, and (b) soil water content and volume of soil air in the four different ecosystems. Closed and open circles respectively signify primary and secondary forests. Open triangles show oil palm plantations; crosses show rubber plantations. The replications were 16 in each ecosystem.

For example, soil temperature explains 73% of seasonal variation in soil respiration in beech and Douglas fir forests (Longdoz et al., 2000). Soil water contents might affect both soil biological activity and physical conditions. Chambers et al. (2004) reported that the relationship between soil respiration rates and soil water contents is curvilinear in an Amazonian tropical forest. Heterotrophic respiration was strongly correlated with soil water contents in a tropical forest, suggesting that the low soil water contents reduced biotic activity, especially that of microbes within the soil (Li et al., 2005). In the present study, however, soil respiration rates displayed a significant and negative correlation with soil water content in the study plot (Figure 4 and Table 1). Results suggest that gas diffusivity of the soil was low under conditions of high soil water contents (Figure 5). This result suggests that soil respiration was affected by the soil's physical condition, especially soil ventilation. Some studies have measured soil gas diffusivity and production using [222]Rn (e.g., Davidson et al., 1995; Uchida et al., 1997).

Table 1. Correlation Coefficient between soil respiration rates and soil physical characteristics

Soil respiration rate (mg CO2 m2 hr-1)		Soil water contents (%)	Volume of gas (%)	Volume of solid (%)	Volume of liquid (%)
Primary forest (n=16)					
Soil respiration rate	1.000				
Soil water content	-0.791	***	1.000			
Volume of gas	0.762	***	-0.789 ***	1.000		
Volume of solid	-0.734	***	0.732 ***	-0.946 ***	1.000	
Volume of liquid	-0.735	***	0.779 ***	-0.974 ***	0.847 ***	1.000
Secondary forest (n=16)					
Soil respiration rate	1.000				
Soil water content	-0.036		1.000			
Volume of gas	0.272		-0.637 **	1.000		
Volume of solid	-0.358		0.223	-0.819 ***	1.000	
Volume of liquid	-0.145		0.799 ***	-0.903 ***	0.494	1.000
Oil palm plantation (n=16)					
Soil respiration rate	1.000				
Soil water content	-0.636		1.000			
Volume of gas	0.586	*	-0.625 **	1.000		
Volume of solid	-0.733	**	0.559 *	-0.914 ***	1.000	
Volume of liquid	-0.050		0.451	-0.688 **	0.334	1.000
Rubber plantation (n=16)					
Soil respiration rate	1.000				
Soil water content	-0.451		1.000			
Volume of gas	0.550	*	-0.570 *	1.000		
Volume of solid	-0.407		0.022	-0.782 ***	1.000	
Volume of liquid	-0.397		0.884 ***	-0.677 **	0.071	1.000

***; $P < 0.0001$, **; $P < 0.01$, *; $P < 0.05$.

Table 2. Soil respiration rate and soil physical characters in the four different ecosystems

Site	Soil respiration rate (mg CO2 m2 hr-1)		Volume of solid (%)		Volume of liquid (%)		Volume of gas (%)
Primary forest	796.0	a	27.8	a	27.5	a	44.7 a
Secondary forest	1176.8	ab	35.6	b	26.5	a	37.9 a
Oil palm plantation	516.6	ac	49.2	c	29.5	a	21.3 b
Rubber plantation	405.8	ac	54.0	c	30.0	a	16.0 b

The different letter among three sites are significantly different (Scheffé's test, $P < 0.05$).

Table 3. Carbon efflux from soil in the world site

Study site	Forest type	Carbon flux form soil (t C ha^{-1} yr^{-1})	Refference
Tropical region			
Brazil	Primary forest	20	Davidson et al.(2000)
Brazil	Secondary forest	18	Davidson et al.(2000)
Brazil	Pasture	15	Davidson et al.(2000)
Costa Rica	Primary forest	10~17	Schwendenmann et al.(2003)
Tailand	Primary forest	26	Hashimoto (2005)
Malaysia (Pasoh)	Primary forest	11	Kira (1987)
Malaysia (Pasoh)	Primary forest	18	The present study
Malaysia (Pasoh)	Secondary forest	18	The present study
Malaysia (Pasoh)	Oil palm plantation	14	The present study
Malaysia (Pasoh)	Rubber plantation	10	The present study
Temperate region			
France	Primary forest (Beech)	7	Epron et al. (1999)
America	Primary forest	5	Toland and Zak (1994)
	Deforestation	5	Toland and Zak (1995)

(a) 1- box model

Litter

Soil organic carbon

(b) 2- box model

Litter

Litter

Soil

(c) 3- box model

Litter

Litter

Soil

Dead root

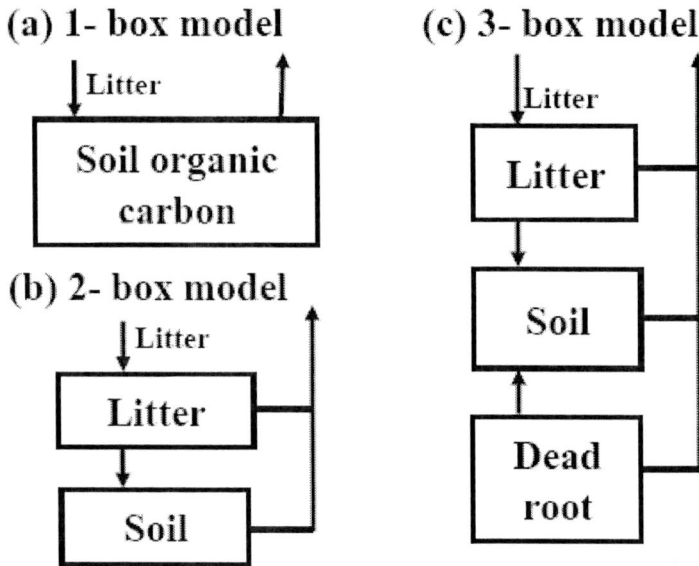

Figure 6. Compartment models of SOC dynamics, as referred from Ito (2002): (a) one-box model, (b) two-box model, (c) three-box model.

The soil respiration rate, as assessed with chamber methods, showed a significant positive correlation with CO_2 efflux from soil using ^{222}Rn (Uchida et al., 1997). This result suggests that soil diffusivity is an important factor for the soil respiration rate.

Annual carbon flux from soil in this study site was compared to results obtained for other ecosystems throughout the world (Table 2). Kosugi et al. (2007) also reported details of spatial and temporal variations in soil respiration, indicating that variation in soil respiration rate increased with plot size. Therefore, our estimate is inferred to be useful for evaluating large areas. Kira (1978) reported annual carbon efflux from soil at a study site, it was 60% of the result found in this study. The results in primary and secondary forests and pasture in Brazil resembled results of this study (Davidson et al., 2000). On the other hand, the results for primary forests in the present study and that of Davidson et al. (2000) were 3–4 times as large as those for cool temperate forests (Toland and Zak, 1994; Epron et al., 1999). In a cool temperate forest, soil temperature is an important factor to decide seasonal variation in soil respiration rates. Therefore, the low soil respiration rate in winter probably affected the annual carbon efflux from soil.

Soil Organic Carbon Dynamics

Soil carbon sequestration was estimated using a one-box model for the primary forest and oil palm plantation, as shown in Table 4. The results suggest that primary forests are a source of carbon. The results of the three-box model suggest that underground biomass is important to estimate the SOC dynamics, especially for tropical forests, which have large aboveground biomass. For example, the aboveground biomass in the primary forest site in this study was 403 t ha^{-1} (Hoshizaki et al., 2004).

Table 4. Assumption of annual SOC storage based on one-box and three-box model in primary forest and oil palm plantation

		Primary forest	Oil palm plantation
	Aboveground biomass (t ha^{-1})	403 a	37.7d
	root biomass (t ha^{-1})	43 b	16 d
	Annual carbon efflux from soil (t C ha^{-1} y^{-1})	18.1 *	14.4 *
H	Heterotrophic respiration (t C ha^{-1} y^{-1})	10.9 (60%) *	6.5 (45%) e
L	Carbon supply from litter (t C ha^{-1} y^{-1})	5.6 c	4.3 e
L - H	Annual soil organic carbon storage by 1-box model (t C ha^{-1})	- 5.3	-2.2
D	dead root biomass (t C ha^{-1} y^{-1})	4.3	1.6
L -H + D	Annual soil organic carbon storage by 3-box model	-1.0	-0.6

a Hoshizaki et al. (2004); b Niiyama and Yamashita unpublished data; CYoneda et al. (2002) the data is the average from 1998   to 2003; d Corley andTinker (2003); e Singh et al. (1999); e the assumed value.

The mean ratio of root biomass and aboveground biomass was reported as ca. 0.21 at these study sites (Niiyama and Yamashita, unpublished data) and 0.25 in a semi-deciduous tropical moist forest in Cameroon (Deans et al., 1996). Therefore, we estimated the root biomass in the primary forest at this study site as about 43 t C ha^{-1}. However, dead root biomass was unknown. In Amazonian tropical forests, dead root biomass is estimated to constitute one-tenth of underground biomass (Malhi and Grace, 2000). Similarly, we estimated that carbon of 4.3 t ha^{-1}, which is 1/10 of 43 t C ha^{-1} is supplied to soil. Consequently, the carbon balance of soil mostly became zero. This result has a large uncertainty. However, it suggests the importance of the amount of dead roots and its turnover time for evaluation of SOC sequestration. Fitter and Hay (2002) suggested that underground biomass tends to increase when soil is in a condition of low soil temperature or soil water content or nutrient suffusion. Therefore, underground biomass and dead root biomass might differ among ecosystems, we must carefully examine these data. In addition, root turnover time must be clarified because it is expected to differ among vegetation types and biomes (Gill and Jackson, 2000). The SOC pooling and sequestration probably differ according to ecosystems, climate change, biomes, and management. Raich et al. (2006) suggested that annual temperature affects SOC because decomposition rates of soil organic matter are influenced by the mean annual temperature. The SOC is higher at higher elevations in wet tropical regions. Annual temperature has a positive correlation with litter fall, the aboveground biomass increment, belowground carbon allocation, and surface fine-litter decay rates (Raich et al., 2006). We must continue to investigate various ecosystems carefully to clarify SOC dynamics and soil sequestration.

ACKNOWLEDGEMENTS

This paper is a part of Global Environment Research Program supported by the Ministry of the Environment, Japan. We thank the Forestry Research Institute Malaysia (FRIM) for giving us permission and E. S. Quah, S. M. Wang, and other staff members for supporting our work in the Pasoh Forest Reserve. We also thank Dr. Okuda and a joint research project of FRIM, University for Pertanian Malaysia (UPM), and the National Institute for Environmental Studies (NIES). We thank Dr. Yashiro of Gifu University for assistance during field investigations.

REFERENCES

Adachi, M., Bekku, Y. S., Kadir, W. R., Okuda, T., Koizumi, H. (2006). Differences in soil respiration between different tropical ecosystems. *Applied Soil Ecology,* 34: 258-265.

Adachi, M., Bekku, Y.S., Konuma, A., Kadir, W. R.., Okuda, T., Koizumi, H. (2005). Required sample size for estimating soil respiration rates in large areas of two tropical forests and of two types of plantation in Malaysia. *Forest Ecology and Management,* 210: 455-459.

Balesdent, J., Chenu, C., Balabane, M. (2000). Relationship of soil organic matter dynamics to physical protection and tillage. *Soil and Tillage Research,* 53: 215-230.

Bekku, Y.S., Kume, A., Masuzawa, T., Kanda, H., Nakatsubo, T., Koizumi, H. (2004). Soil respiration in a high arctic glacier foreland in Ny-Ålesund, Svalbard. *Polar Bioscience,* 17:36-46.

Chambers, J.Q, Tribuzy, E.S, Toledo, L.C, Crispim, B.F, Higuchi, N., Santos, J.D., Araújo, A.C., Kruijt, B., Nobre, A.D., Trumbore, S.E. (2004). Respiration from a tropical forest ecosystem: partitioning of sources and low carbon use efficiency. *Ecol. Appl.* 14: S72-S88.

Cleveland, C. C., Nemergut, D.R., Schmidt, S.K., Townsend, A.R. (2007). Increases in soil respiration following labile carbon additions linked to rapid shifts in soil microbial community composition. *Biogeochemistry* 82: 229-240.

Corley, R.H.V., Tinker, P.B., editors. (2003). *The oil palm.*Oxford, Blackwell Science.

Davidson, E.A., Trumbore, S.E. (1995). Gas diffusivity and production of CO_2 in deep soil of the eastern Amazon. *Tellus* 47B: 550-565.

Davidson, E.A., Verchot, L.V., Cattanio, J.H., Ackerman, I.L., Carvaho, J.E.M. (2000). Effect of soil water content on soil respiration in forests and cattle pastures of eastern Amazonia. *Biogeochemistry* 48: 53–69.

Deans, J.D., Moran, J., Grace, J. (1996). Biomass relationships for tree species in regenerating semi-deciduous tropical moist forest in Cameroon. *Forest Ecology and Management* 88: 215-225

Elberling B. (2007). Annual soil CO_2 efflux in the High Arctic: The role of snow thickness and vegetation types. *Soil Biology and Biochemistry* 39: 646-654.

Epron, D., Farque, L., Lucot, E., Badot, P.M. (1999). Soil CO_2 efflux in a beech forest: dependence on soil temperature and soil water content. *Annals forest Science* 56: 221–226.

Epron, D., Nouvellon, Y., Deleporte, P., Ifo, S., Kazotti, G.., M'Bou, A.T., Mouvondy, W., Saint-André, L, Roupsard, O., Jourdan, C., Hamel, O. (2006). Soil carbon balance in a clonal Eucalyptus plantation in Congo: effect of logging on carbon input and soil CO_2 efflux. *Global Change Biology* 12:1021-1031.

Epron, D., Nouvellon, Y., Roupsard, O., Mouvondy, W., Mabiala, A., Saint-André, L., Joffre, R., Jourdan, C., Bonnefond, J.M., Berbigier, P., Hamel, O. (2004). Spatial and temporal variations of soil respiration in a Eucalyptus plantation in Congo. *Forest Ecology and Management* 202: 149-160.

Fang, C., Moncrieff, J.B., Gholz, H.L., Clark, K.L. (1998). Soil CO_2 efflux and its spatial variation in a Florida slash pine plantation. *Plant and Soil* 205: 135–146.

Field, C.B., Behrenfeld, M.J., Randerson, J.T., Falkowski, P. (1998). Primary production of the biosphere: integrating terrestrial and oceanic components. *Science* 281: 237–240.

Fitter, A.H., Hay, R.K.M., editor. (2002). *Environmental Physiology of Plants*. Third edition. California: Academic press.

Gil,l R.A., Jackson, R.B. (2000). Global patterns of root turnover for terrestrial ecosystems. *New Phytologist* 147: 13-31.

Golchin, A., Oades, J.M., Skjemstad, J.O., Clarke, P. 1994. Soil structure and carbon cycling. *Australian Journal of Soil Research* 32: 1043-1068.

Hashimoto, S. (2005). Temperature sensitivity of soil CO_2 production in a tropical hill evergreen forest in northern Thailand. *Journal of Forest Research*. 10: 497-503.

Hillel. D, editor. (1998). *Environmental soil physics*. California: Elsevier.

Hoshizaki, H., Niiyama, K., Kimura, K., Yamashita, T., Bekku, Y., Okuda, T., Quah, E.S., Nur Supardi, M.N. (2004). Recent biomass change in relation to stand dynamics in a lowlamd tropical rain forest, Pasoh Forest Reserve, Malaysia. *Ecological Research,* 19: 357-363.

Houghton, R.A., Hackler, J.L. (1999). Emissions of carbon from forestry and land-use change in tropical Asia. *Global Change Biology,* 5: 481-492.

Houghton, R.A. (2003). Revised estimates of the annual net flux of C to the atmosphere from changes in land use and land management 1850-2000. *Tellus,* 55B: 378-390.

Ito, A. (2002). Soil organic carbon storage as a function of the terrestrial ecosystem with respect to the global carbon cycle. *Japanese Journal of Ecology,* 52: 189-227

Jobbagy, E.G., Jacksonm, R.B. (2000). The vertical distribution of soil organic carbon and its relation to climate and vegetation. *Ecological Applications,* 10 (2): 423-436.

Kiese, R., Butterbach-bahl, K. (2002). N_2O and CO_2 emissions from three different tropical forest sites in the wet tropics of Queensland, Australia. *Soil Biology and Biochemistry* 34: 975–987.

Kira, T. (1987). Primary production and carbon cycling in a primeval lowland rainforest of peninsular Malaysia. *Tree Crop Physiology* :99-119.

Kosugi, Y., Mitani, T., Itoh, M., Noguchi, S., Tani, M., Matsuo, N., Takanashi, S., Ohkubo, S., Abdul Rahim Nik. (2007). Satial and temporal variation in soil respiration in a southeast Asian tropical rain forest. *Agricultural and Forest Meteorology* 147: 35-47.

Kursar, T.A. (1989). Evaluation of soil respiration and soil CO_2 concentration in a lowland moist forest in Panama. *Plant and Soil* 113: 21–29.

Lal, R. (2004). Soil Carbon Sequestration Impacts on Global Climate Change and Food Security. *Science* 304, 1263-1627.

Li, Y., Xu, M., Zou, X., Xia, Y. (2005). Soil CO_2 efflux and fungal and bacterial biomass in a plantation and a secondary forest in wet tropics in Puerto Rico. *Plant and Soil* 268: 151-160.

Longdoz, B., Yernaux, M., Aubinet, M. (2000). Soil CO_2 efflux measurements in a mixed forest: impact of chamber disturbances, spatial variability and seasonal evolution. *Global Change Biology* 6: 907-917.

Malhi, Y., Baldocchi, D.D., Jarvis, P.G.. (1999). The carbon balance of tropical, temperate and boreal forests. *Plant, Cell and Environment* 22: 715-740.

Malhi, Y., Grace, J. (2000). Tropical forests and atmospheric carbon dioxide. *Tree* 15(8): 332–337.

Mo, W., Lee, M., Uchida, M., Inatomi, M., Saigusa, N., Mariko, S., Koizumi, H. (2005). Seasonal and annual variations in soil respiration in a cool temperate deciduous broad-leaved forest in Japan. *Agric. For. Meteorol.* 134: 81-94.

Neergaard, A.D., Porter, J.R., Gorissen, A. (2002). Distribution of assimilated carbon in plants and rhizosphere soil of basket willow (Salix viminalis L.) *Plant and Soil* 245:307-314.

Nishimura, S., Yonemura, S., Sawamoto, T., Shirato, Y., Akiyama, H., Sudo, S., Yagi, K. (2008). Effect of land use change from paddy rice cultivation to upland crop cultivation on soil carbon budget of a cropland in Japan. *Agriculture Ecosystems and Environment* 125: 9-20.

Ohashi, M., Kume, T., Yamane, S., Suzuki, M. (2007). Hot spots of soil respiration in an Asian tropical rainforest. *Geophysical Research Letters* 34:L08705.

Okuda, T., Suzuki, M., Adachi, N., Quah,, E.S., Hussein, N.A., Manokaran, N. (2003). Effect of selective logging on canopy and stand structure and tree species composition in a lowland dipterocarp forest in peninsular Malaysia. *Forest Ecology and Management* 175: 297-320.

Raich, J.W., Russell, A.E., Kitayama, K., Parton, W.J., Vitousek, P.M. (2006). Temperature influences carbon accumulation in moist tropical forests. *Ecology* 87(1): 76-87.

Raich, J.W., Tufekcioglu, A. (2000). Vegetation and soil respiration: correlations and controls. *Biogeochemistry* 48: 70-90.

Raich, J.W., Schlesinger, W.H. (1992). The global carbon dioxide flux in soil respiration and its relationship to vegetation and climate. *Tellus* 44B: 81-99

Rayment, M.B., Jarvis, P.G.. (2000). Temporal and spatial variation of soil CO_2 efflux in a Canadian boreal forest. *Soil Biology and Biochamistry* 32: 35-45.

Schwendenmann, L., Veldkamp, E. (2006). Long-term CO_2 production from deeply weathered soil of a tropical rain forest: evidence for a potential positive feedback to climate warming. *Global Change Biology* 12: 1878-1893.

Schwendenmann, L., Veldkamp, E., Brenes, T., O'Bien, J., Mackensen, J. (2003). Spatial and temporal variation in soil CO_2 efflux in an old-growth neotropical rain forest, La Selva, Costa Rica. *Biogeochemistry*, 64: 111-128

Singh, J.S., Gupta, S.R. (1977). Plant decomposition and soil respiration in terrestrial ecosystems. *Botanical Review* 43: 449–528.

Stoyan, H., De-Polli, H., Böhm, S., Robertson, G.P., Paul, E.A. (2000). Spatial heterogeneity of soil respiration and related properties at the plant scale. *Plant and Soil* 222: 203-214.

Sotta, E. D., Meir, P., Malhi, Y., Nober, A. D., Hodnett, M. and Grace, J. 2004. Soil CO_2 efflux in a tropical forest in the central Amazon. *Global Change Biology* 10, 601-617.

Søe, A.R.B., Buchmann, N. (2005). Spatial and temporal variations in soil respiration in relation to stand structure and soil parameters in an unmanaged beech forest. *Tree Physiol.* 25: 1427-1436.

Tang, Y., Kachi, N., Furukawa, A., Muhamad, A. (1996). Light reduction by regional haze and its effect on simulated leaf photosynthesis in a tropical forest of Malaysia. *Forest Ecology and Management* 89: 205–211.

Toland, D.E., Zak, D.R. (1994). Seasonal patterns of soil respiration in intact and clear-cut northern hardwood forests. *Canadian Journal of Forest Research* 24: 1711–1716.

Uchida, M., Nojiri, Y., Saigusa, N., Oikawa, T. (1997). Calculation of CO_2 flux from forest soil using [222]Rn calibrated method. *Agricultural and Forest Meteorology* 87: 301-311.

Xu, M, Qi Y. (2001). Soil-surface CO_2 efflux and its spatial and temporal variation in a young ponderosa pine plantation in northern California. *Global Change Biology* 7: 667–677.

Yamashita, T., Kasuya, N., Wan Rasidah, K., Suhaimi Wan, C., Quah, E.S., Okuda, T. (2003). Soil and belowground characteristics of Pasoh Forest Reserve. Okuda T., Manokaran Y, Matsumoto Y, Niiyama K, Thomas SC, Ashton PS, editors. *Pasoh, Ecology of a lowland rain forest in Southeast Asia.* Springer-Verlag, Tokyo. p89-109.

Yashiro, Y., Kadir, W.R., Adachi, M., Okuda, T., Koizumi, H. (2007). Emission of nitrous oxide from tropical forest and plantation soils in Penincular Malaysia. *Tropics* 17:17-23.

Yoneda, T., Kimura, K., Numata, S., Okuda, T., Niiyama, K., Kadir, W.R., (2002). Temporal and spatial distributions of litter-fall rates in Pasoh Forest Reserve with special references to fine litter fall. *Research report of the NIES/ FRIM/ UPM Joint Research Project.* p.16-29.

In: Forest Canopies: Forest Production, Ecosystem…   ISBN 978-1-60741-457-5
Editor: J. D. Creighton and P. J. Roney   © 2009 Nova Science Publishers, Inc.

Chapter 6

CARBON STABLE ISOTOPES OF MAMMAL BONES AS TRACERS OF CANOPY DEVELOPMENT AND HABITAT USE IN TEMPERATE AND BOREAL CONTEXTS

Dorothée G. Drucker[1] and Hervé Bocherens[2]

1.Institut für Ur- und Frühgeschichte, Naturwissenschaftliche Archäologie,
Universität Tübingen, Germany;
2.Institut für Geowissenschaften, Biogeologie,
Universität Tübingen, Germany

ABSTRACT

Plants growing under dense canopy experience conditions - low light intensity, recycling of biogenic CO_2 - that lead to lower $^{13}C/^{12}C$ ratios than plants growing in open landscapes. This particular stable isotopic ratio is passed on tissues of herbivorous mammals feeding under dense canopy conditions, including their bone collagen. As bone collagen can be extracted from ancient bones several thousand years old, its carbon isotopic signature can be used to track the development of dense forest through time. One good example is the development of dense forest in France during the last 35,000 years, especially since the beginning of the Holocene about 10,000 years ago. Using dated large bovid bones and teeth (from Bison *Bison priscus* and Aurochs *Bos primigenius*) from archaeological sites as tracers of dense forest development, it appears that open conditions dominated from 35,000 to 10,000 years ago, although changes in the vegetation composition were observed. In contrast, a very significant decrease of $^{13}C/^{12}C$ ratio during the Preboreal period, around 10,000 years ago, seems to correspond to the spread of dense temperate forest in France and is directly linked to climatic change. A further decrease of $^{13}C/^{12}C$ ratio in aurochs bone collagen occurs during the Late Atlantic (6,000-5,000 years ago), which coincides with the extension of agricultural societies and the development of cattle husbandry. This further decrease seems to correspond to some clearance of the dense forest cover for agricultural purpose and particularly to provide feeding grounds for cattle, while the wild aurochs tend to live in deep dense forest and

---
[1] e-mail: dorothee.drucker@ifu.uni-tuebingen.de, address: Rümelinstrasse 23, D-72070 Tübingen, Germany.
[2] e-mail: herve.bocherens@uni-tuebingen.de, address: Sigwartstrasse 10, D-72076 Tübingen, Germany.

avoid forest edges and open environments. When local situations are explored further, for instance the Paris area and the French Jura during the Middle Neolithic, regional differences are found regarding the use of dense forest biomass by prehistoric farmers. This new approach is a promising one that could allow fine level tracking of dense forest exploitation by humans through time.

INTRODUCTION

In tropical forests, atmospheric CO_2 available to plants in poorly ventilated understory is ^{13}C depleted relative to the general atmosphere as the result of CO_2 recycling from leaf litter (e.g. van der Merwe and Medina, 1991). In addition, a CO_2 concentration gradient and light attenuation under the forest canopy lead to depleted ^{13}C abundances in understory plants due to change in photosynthetic activity and stomatal conductance (e.g. Francey et al., 1985; Broadmeadow et al., 1992). Based on similar mechanisms, plants growing under the canopy of boreal and temperate forests also exhibit ^{13}C depletion compared to the same plant groups in open conditions (e.g., Vogel, 1978; Schleser and Jayasekera, 1985; Brooks et al. 1997).

This particular stable isotopic ratio is passed on tissues of herbivorous mammals feeding under dense canopy conditions, including their bone collagen. Until recently, the influence of canopy effect on the ^{13}C abundance in herbivores from boreal and temperate forest had barely been examined. A recent study of modern populations of ungulates, including reindeer, red deer, roe deer and bison, clearly showed that individuals dwelling under dense forest conditions exhibited significantly lower ^{13}C amounts than individuals of the same species dwelling under open environmental conditions (Drucker et al., 2008).

An illustrative example is provided by the case of roe deer from the forest of Dourdan, near Paris in France, where some individuals lived deep in the forest, while others lived close to the forest edge. Plants from the forest understory present $\delta^{13}C$[3] values significantly lower than plants of cultivated fields of wheat and rape nearby. The roe deer belonging to the two groups - forest-interior versus forest-edge - clearly follow the trend of their dietary input (Figure 1). Carbon isotopic signatures allow in such case to quantify the reliance of each individual of a game species on forest dietary resources. It therefore permits to evaluate the proportion of wild animals that depends directly on forest resources for its feeding behaviour.

Such approach is possible for ancient mammals as well. Indeed, bone collagen can be extracted from ancient bones several thousand years old (e.g. Bocherens et al., 1999; Drucker et al., 2003). Its carbon isotopic signature can be used to track the development of dense forest through time in a given region, and to track the differential use of closed forested habitat by different species in a given ancient environment.

---

[3] Isotopic abundances are expressed as δ values as follows: $\delta^{13}C = (^{13}C/^{12}C_{sample}/^{13}C/^{12}C_{standard}-1)*1000$ (‰). The international reference standard is Vienna-PeeDee Belemnite (V-PDB) for $\delta^{13}C$ measurements.

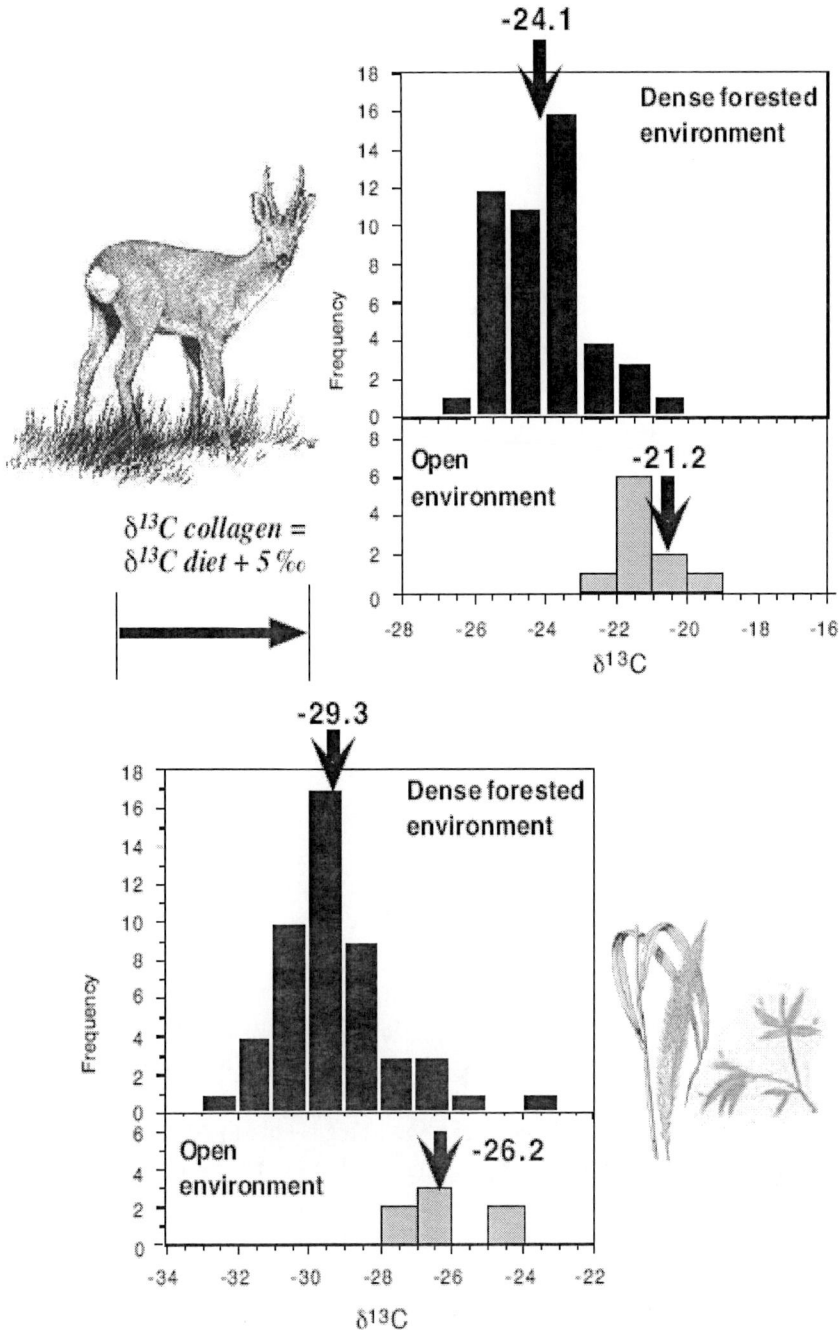

Figure 1. Carbon isotopic abundances ($\delta^{13}C$) of plants from the Dourdan forest and the crop plants nearby (below) and of roe deer collagen from individuals living deep in the forest and close to the forest edge (above).

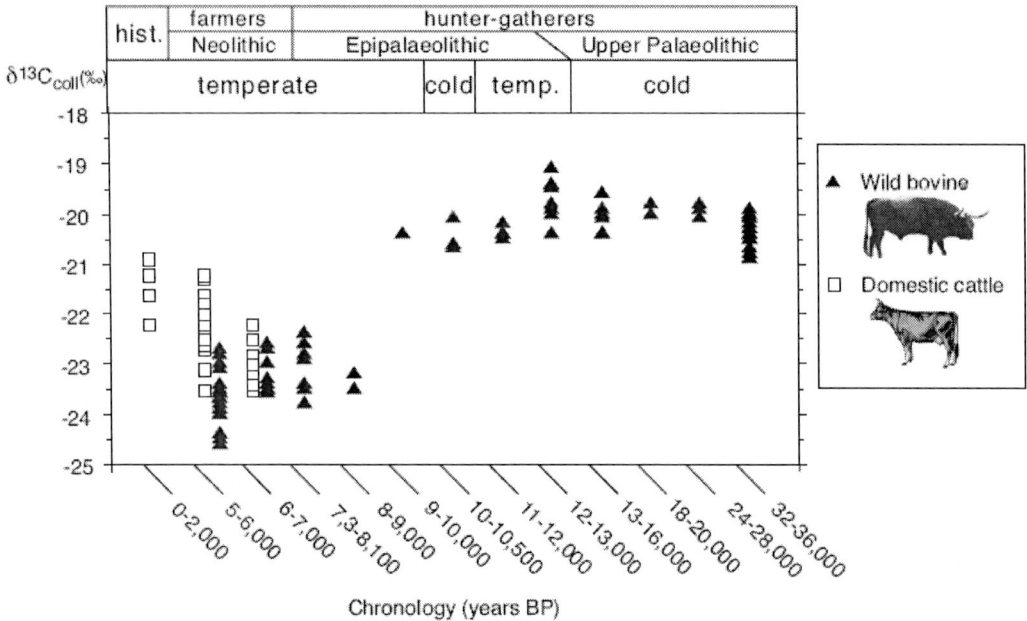

Figure 2. Carbon isotopic abundances (δ^{13}C) of large bovid collagen through time in France.

TRACKING DEVELOPMENT OF DENSE
FOREST THROUGH TIME IN FRANCE

One good example is the development of dense forest in France during the last 35,000 years, especially since the beginning of the Holocene about 10,000 years ago (Figure 2). Using dated large bovid bones and teeth (from steppe bison *Bison priscus* and aurochs *Bos primigenius*) from archaeological sites as tracers of dense forest development, it appears that open conditions dominated from 35,000 to 10,000 years ago, although changes in the vegetation composition were observed. In contrast, a very significant decrease of ^{13}C/^{12}C ratio occurs probably during the Preboreal around 10,000 years ago and is clearly established during the Boreal period, around 9000 years ago. This seems to correspond to the spread of dense temperate forest in France and is directly linked to climatic change. The oldest domestic cattle are in the same range of δ^{13}C values as the wild aurochs, which seems to correspond to a similar forested habitat. A further decrease of ^{13}C/^{12}C ratio in aurochs bone collagen occurs during the Late Atlantic, while domestic cattle exhibits an increase of their ^{13}C/^{12}C ratio. This coincides with the extension of agricultural societies and the development of cattle husbandry. This phenomenon can be related to some clearance of the dense forest cover for agricultural purpose and particularly to provide feeding grounds for cattle, while the wild aurochs tend to take refuge in deep dense forest and avoid forest edges and open environments. A similar isotopic pattern reflecting understory plants browsing by wild bovines and open area grass consumption by early domestic cattle has been documented in Denmark (Noe-Nyggard et al., 2005). The habitat partitioning between domestic and wild bovines could have been a constant in Western Europe. The dependence of European wild bovine on forest resources appears as a recent event, which probably amplified with human pressure.

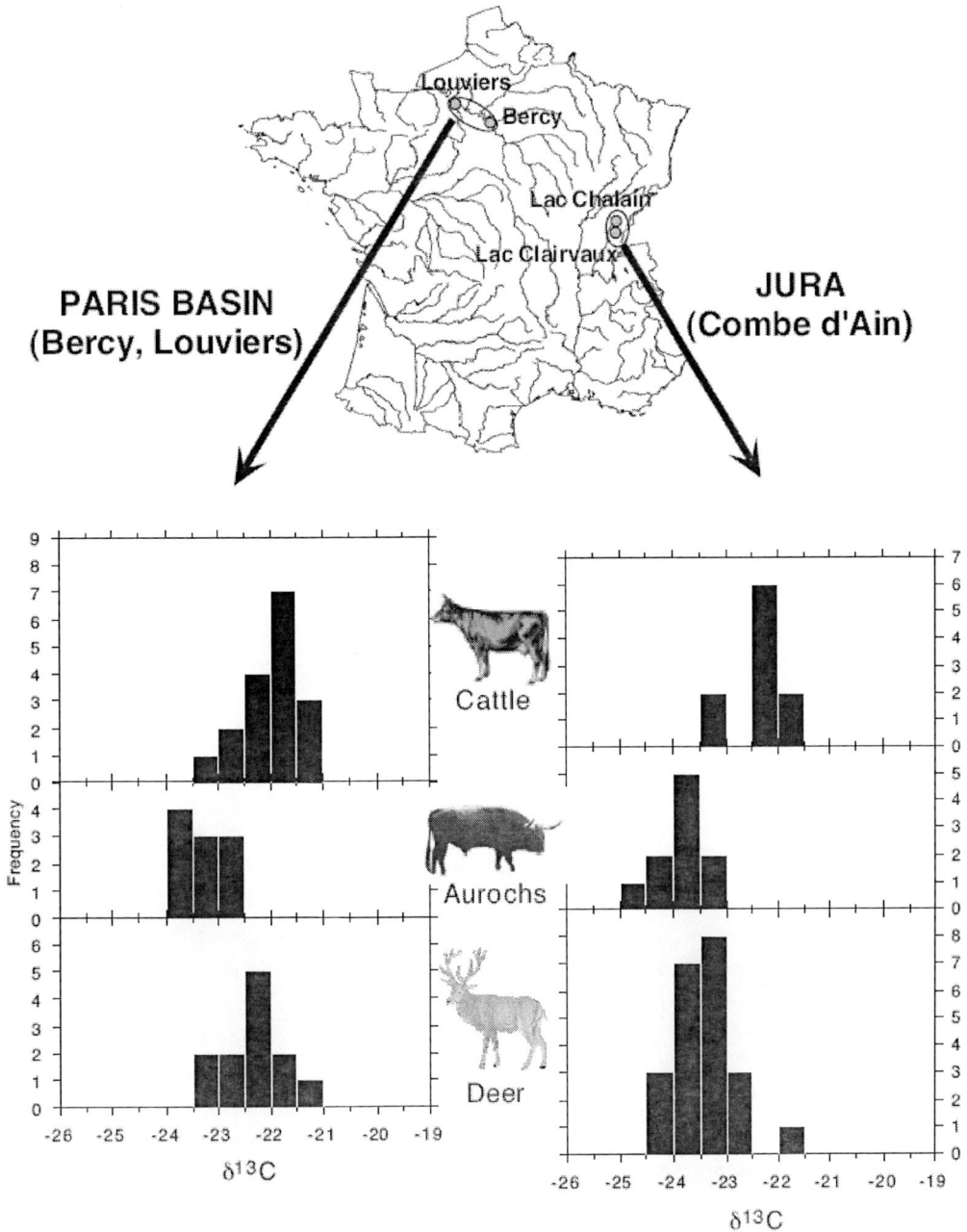

Figure 3. Location of the middle Neolithic sites from Paris Basin and Jura and $\delta^{13}C$ values of domestic cattle, wild aurochs and red deer in both regions.

TRACKING THE DIFFERENTIAL USE OF
DENSE FORESTED HABITAT DURING THE MIDDLE NEOLITHIC IN
PARIS BASIN AND JURA (FRANCE)

When local situations are explored further, regional differences are found regarding the use of dense forest biomass by prehistoric farmers. For instance, some isotopic analyses were performed on bones from domestic cattle, wild aurochs and red deer from early agricultural sites (Neolithic) dated from around 3,000 to 4,000 years BC, located in the Paris Basin (Louviers and Bercy) and in the French Jura (Lac Chalain and Lac Clairvaux) (Pétrequin, 1996; Bocherens et al., 1997). The results from both sites from the Paris Basin are similar for each species and therefore have been plotted together. Sites from the French Jura are also considered together as no significant difference is detectable for a given species. In the two regions, it appears clearly that domestic cattle exhibit significantly less negative $\delta^{13}C$ values than wild aurochs (Figure 3). This reflects the dense forest habitat of the wild aurochs, while domestic cattle dwelled in more open pastures. Furthermore, comparison between the two regions shows that each species presents more negative $\delta^{13}C$ values in the Jura sites than in the Paris Basin sites. The more negative $\delta^{13}C$ values of aurochs in Jura are interpreted as a consequence of a denser forest around the sites of the Jura, which is confirmed by the isotopic results obtained on red deer. It is noteworthy that red deer hunting was more developed in Jura than in the Paris Basin (Arbogast and Pétrequin, 1993; Giligny, 2005). This example shows that prehistoric farmers from Jura used more intensely the dense forested biomass than those from the Paris Basin during the same period.

CONCLUSION

This new approach is a promising one that could allow fine level tracking of dense forest exploitation by humans and large herbivores through time. Carbon stable isotopes study allows deciphering the degree of dependence of a herbivorous species to closed forested habitat according to time and space. The importance of forest resources can also be estimated for predators, using the carbon isotopic signatures of hunted species or directly using the carbon isotopic values of predators. For example, a Neandertal specimen from Belgium that was found together with herbivores dwelling in dense forest according to their carbon isotopic signatures did not show the same ^{13}C-depletion, indicating that its prey came essentially from open environments (Bocherens et al., 1999). This approach can also be used for modern ecosystems, to compare the actual use of dense forest resources by different ecotypes of the same species, as it was demonstrated on reindeer (Drucker, 2007).

REFERENCES

Arbogast, R.-M. and Pétrequin, P. (1993). La chasse au cerf au Néolithique dans le Jura: gestion d'une population animale sauvage. In *Exploitation des animaux sauvages à travers le temps*, (XIIIe Rencontres internationales d'Archéologie et d'Histoire d'Antibes, pp. 221-232). Antibes: Editions APDCA.

Bocherens, H., Tresset, A., Wiedemann, F., Giligny, F., Lafage, F., Lanchon, Y. and Mariotti, A. (1997). Bone diagenetic evolution in two French Neolithic sites. *Bulletin de la Société Géologique de France, 168,* 555-564.

Bocherens, H., Billiou, D., Patou-Mathis, M., Otte, M., Bonjean, D., Toussaint, M. and Mariotti, A. (1999). Palaeoenvironmental and palaeodietary implications of isotopic biogeochemistry of late interglacial Neandertal and mammal bones in Scladina Cave (Belgium). *Journal of Archaeological Science, 26,* 599-607.

Broadmeadow, M.S.J., Griffiths, H., Maxwell, C. and Borland, A.M. (1992). The carbon isotope ratio of plant organic material reflects temporal and spatial variations in CO_2 within tropical forest formations in Trinidad. *Oecologia, 89,* 435-441.

Brooks, J.R., Flanagan, L.B., Buchmann, N. and Ehleringer, J.R. (1997). Carbon isotope composition of boreal plants: functional grouping of life forms. *Oecologia, 110,* 301-311.

Drucker, D.G. (2007). Les cervidés durant le Tardiglaciaire et l'Holocène ancien en Europe occidentale : approche isotopique. In S. Beyries, and V. Vaté (Eds.), *Les civilisations du renne d'hier et d'aujourd'hui: Approches ethnohistoriques, archéologiques et anthropologiques*, (XXVIIe rencontres internationales d'archéologie et d'histoire d'Antibes, pp. 243-253). Antibes: Editions APDCA.

Drucker, D. G., Bocherens, H. and Billiou, D. (2003). Evidence for shifting environmental conditions in Southwestern France from 33,000 to 15,000 years ago derived from carbon-13 and nitrogen-15 natural abundances in collagen of large herbivores. *Earth and Planetary Science Letters, 216,* 163-173

Drucker, D. G., Bridault, A., Hobson, K. A., Szuma, E. and Bocherens, H. (2008). Can carbon-13 abundances in large herbivores track canopy effect in temperate and boreal ecosystems? Evidence from modern and ancient ungulates. *Palaeogeography, Palaeoclimatology, Palaeoecology, 266,* 69-82.

Francey, R.J., Gifford, R.M., Sharkey, T.D. and Weir, B. (1985). Physiological influences on carbon isotope discrimination in huon pine (*Lagarostrobos franklinii*). *Oecologia, 66,* 211-218.

Giligny, F. (Ed.) 2005. Louviers "La Villette" (Eure): un site néolithique moyen en zone humide. *Documents Archéologiques de l'Ouest.* 343p.

Noe-Nygaard, N., Price, T.D. and Hede, S.U. (2005). Diet of aurochs and early cattle in southern Scandinavia: evidence from ^{15}N and ^{13}C stable isotopes. *Journal of Archaeological Science, 32,* 855-871.

Pétrequin, P. (1996). Management of architectural woods and variations in population density in the fourth and third millenia B.C. (Lakes Chalain and Clairvaux, Jura, France). *Journal of Anthropological Archaeology, 15,* 1-19.

Schleser, G.H. and Jayasekera, R. (1985). $\delta^{13}C$-variations of leaves in forests as an indication of reassimilated CO_2 from the soil. *Oecologia, 65,* 536-542.

van der Merwe, N.J. and Medina, E. (1991). The canopy effect, carbon isotope ratios and foodwebs in Amazonia. *Journal of Archaeological Science, 18,* 249-259.

Vogel, J.C. (1978). Recycling of carbon in a forest environment. *Œcologia Plantarum, 13,* 89-94.

In: Forest Canopies: Forest Production, Ecosystem...   ISBN 978-1-60741-457-5
Editor: J. D. Creighton and P. J. Roney   © 2009 Nova Science Publishers, Inc.

Chapter 7

SIMULATING THE TWO-WAY FEEDBACK BETWEEN TERRESTRIAL ECOSYSTEMS AND CLIMATE: IMPORTANCE OF FOREST ECOLOGICAL PROCESSES ON GLOBAL CHANGE

Takeshi Ise[1], Tomohiro Hajima, Hisashi Sato and Tomomichi Kato

Frontier Research Center for Global Change, Japan Agency
for Marine-Earth Science and Technology,
Yokohama, 236-0001, Japan

ABSTRACT

The ongoing anthropogenic climate change is immensely altering structure and function of the terrestrial biosphere, including forest ecosystems. In turn, the changing ecosystems have a strong potential to modify the climate through changes in biogeochemical cycles (e.g., C storage) and biophysics (e.g., albedo and hydrological cycling). Forest ecosystems will have particularly significant impacts onto the climate due to their large terrestrial coverage, vast C stock, and prominent biophysical characteristics. To reproduce the two-way interaction between vegetation and climate, climate models should be integrated with dynamically responding vegetation models. Here we present our recent progress, concerns, and future directions in simulations of vegetation processes by the terrestrial biosphere model (TBM) sSEIB (a simplified version of SEIB-DGVM) that is coupled to a climate system model (Center for Climate System Research-Frontier Research Center for Global Change general circulation model, CCSR-FRCGC GCM). sSEIB explicitly reproduces the ecophysiological, population, and community dynamics based on an individual-based forest model representation. The model is also fully coupled to the global biogeochemical cycling that in turn affects atmospheric CO_2 concentrations. The GCM-coupled sSEIB successfully reproduced the current global distributions of vegetation types and plant production. A preliminary climate change experiment with the stand-alone sSEIB showed significant responses of terrestrial vegetation and soil C storage.

1 author of correspondence: ise@jamstec.go.jp; +81-45-778-5595.

INTRODUCTION

We ecologists have long known that the climate is the essential determinant of vegetation (biogeography; e.g., Whittaker 1975). However, we also note that the vegetation—the dominant characteristic of the land surface—strongly affects the climate (e.g., Charney 1975; Betts et al. 1997; Levis et al. 1999; Foley et al. 2003; Luo 2007; Bonan 2008A). In the age of climate change, we have to study this two-way interaction between climate and vegetation in order to make a realistic projection of the future conditions in an integrative manner (Moorcroft 2003). Although the effects of vegetation onto the climate are significant, however, the direction and magnitude of the land-atmosphere feedback is only loosely constrained in the current generation of climate models (IPCC 2007).

Impacts of terrestrial ecosystems onto the climate can be summarized into two categories. The first mechanism is (1) biophysical processes of vegetation. For instance, the change in vegetation can alter albedo, or surface reflectance. Due to the climate change, for example, if tree species colonize tundra areas, the regional albedo will be significantly lowered and more fractions of solar radiation will be absorbed. This loop mediated by the surface albedo change will have a positive feedback on land and atmospheric changes because it intensifies the climatic warming (Bonan et al. 2003). Another important biophysical effect is the surface water balance. Transpiration, one of the most essential plant physiological processes, is strongly controlled by stomatal conductance that sensitively responds to environmental conditions. Evaporation is also a function of vegetation cover since interception by plant canopy increases evaporative loss and decreases water infiltration into soil. Evapotranspiration—the rate of water vapor from land surface—affects atmospheric circulation, such as precipitation patterns (Charney 1975). Moreover, changes in energy balance such as the Bowen ratio are deeply influenced by vegetation, and the resultant regional climate is modified. These biophysical processes are mainly short-term factors that are primarily caused by ecophysiological responses of vegetation, and have strong seasonality such as spring/fall leaf flush/drop.

Another important mechanism of terrestrial feedback is (2) biogeochemical processes such as C cycling. The atmospheric CO_2 concentrations are strongly regulated by terrestrial ecosystems—through the balance between photosynthesis and respiration and changes in ecosystem C stocks. Among the variety of vegetation types, forests are an especially important vegetation type because they accumulate a large portion of the assimilated C in biomass. The biomass storage of C is a relatively long-term process, and the accumulated biomass is not a simple function of the short-term ecophysiological parameters but a complex outcome of long-term ecological processes such as tree growth and mortality. The forest composition—species, age, and size—is also affected by intra- and inter-specific competition. To explicitly simulate population and community ecology, a height-structured, individual-based forest model is thought to be an appropriate tool to reproduce and project the land-atmosphere feedbacks in an integrated manner (Purves and Pacala 2008). In addition, vegetation also controls the C input to the soil organic matter—another large reserve of C.

Many recent GCMs are coupled with terrestrial biosphere models, or TBMs (e.g., GFDL; Delworth et al. 2006). Most of these TBMs are so called "big leaf" models as they mainly simulate short-term ecophysiological processes by assuming large GCM grids are covered by homogeneous vegetation. The long-term ecological processes such as biomass accumulation, competition, survival, and mortality, are often simplified by empirical relationships (e.g., TRIFFID; Cox 2001). As a result, big-leaf TBMs may not be able reproduce complex vegetation response to environmental changes in a long-term. For example, tree growth is largely affected by local, gap-scale ecological interactions as well as the large-scale climatic factors (Ise and Moorcroft 2008). The response of vegetation may show a delay after the climatic forcing, and this time lag can in turn affect the trajectory of climate change. For example, the warming in the high-latitudinal regions can stimulate colonization of tree species on tundra, and this vegetation shift will have a large effect on climate due to changes in biophysics and biogeochemistry, as we mentioned earlier. In this scenario, the rate and extent of this colonization process is dependent on subgrid-scale ecological characteristics, such as seed dispersal, seedling establishment, growth, survival, and competition. Moreover, these local ecological mechanisms may cause other nonlinear responses of the biosphere. For example, especially when adult trees dominate the vegetation, the terrestrial ecosystem often shows a resistance against environmental stress factors. The accumulation of recalcitrant soil organic carbon (SOC) is another controlling factor of the long-term ecological memory.

In this chapter, we report the progress from our current project on the simulation of global terrestrial vegetation that is designed to be fully coupled to a climate model. Our global dynamic vegetation model sSEIB (a simplified version of SEIB-DGVM; spatially explicit, individual-based dynamic global vegetation model; Sato et al. 2007) explicitly captures local, gap-scale processes forest ecology and translate them to the global ecosystem function. By explicitly simulating population and community dynamics among individual trees, the model has a capacity to reproduce transient vegetation changes (e.g., time lag) in ecological timescales. The vegetation dynamics reproduced in sSEIB responds to environmental changes (e.g., air temperature, precipitation, and CO_2 concentrations) simulated by the climate model, and the ecosystem variables (e.g., carbon storage and leaf area index—LAI) affect the atmospheric dynamics and radiative forcing.

MODEL DESCRIPTION

sSEIB is a dynamically responding TBM designed to be coupled to CCSR-FRCGC GCM (Kawamiya et al. 2005) and the land-surface model MATSIRO (Minimal Advanced Treatments of Surface Interaction and RunOff; Takata et al. 2003), which connects atmospheric processes and biological processes by simulating land-surface biophysics (Figure 1). Based on SEIB-DGVM (Sato et al. 2007), sSEIB is an individual-based model of forest structure and function, and ecosystem processes (e.g., biogeochemical cycling) are fully coupled to vegetation dynamics. Photosynthesis and autotrophic respiration are calculated from process-based ecophysiological submodels. Tree growth, population dynamics, and community processes are simulated by an individual-based approach (*sensu* SORTIE (Pacala et al. 1993), FORET (Shugart 1984), and JABOWA (Botkin et al. 1972)). Biogeochemical processes of the terrestrial biosphere are integrated to vegetation ecology, based on Sim-CYCLE (Ito and Oikawa 2002) and LPJ DGVM (Sitch et al. 2003). Here, we briefly describe some characteristics of sSEIB. More details of ecological dynamics are found in Sato et al. (2007).

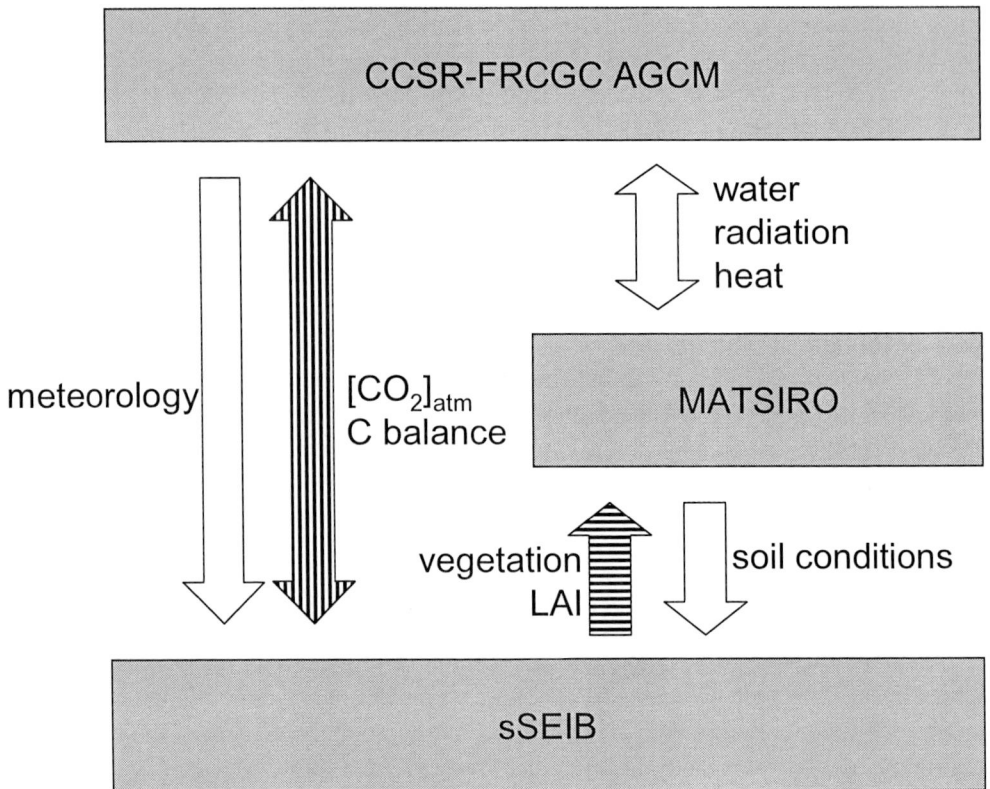

Figure 1. Schematic diagram of the coupling among the dynamic global vegetation model (sSEIB), the GCM (CCSR-FRCGC), and the land-surface model (MATSIRO). The arrow with vertical stripes represents the biogeochemical feedback between terrestrial ecosystem and the atmosphere. The arrow with horizontal stripes represents the biophysical feedback, and this process affects the climate through land-surface dynamics.

Plant Functional Types (PFTS)

To efficiently simulate the large global gradient of ecological characteristics, the biodiversity in plant species is aggregated in several plant functional types, or PFTs (Table 1). Ecophysiological and population dynamics parameters are obtained from SEIB-DGVM (Sato et al. 2007). Each PFT has different ecophysiological parameters such as maximum photosynthetic rates, optimal temperatures for photosynthesis, and minimum temperatures for frost-related mortality. Growth rates and C allocation also differ, resulting in differential growth pattern and interspecific competition among PFTs based on the environmental conditions of a GCM grid cell. Thus, sSEIB has a capacity to reproduce the structure and composition of the natural vegetation arisen from the climate regime, and the resultant biome types can shift as the climate changes. Note that tropical broadleaf evergreen PFTs (PFTs = 1, 2, 3, 4) have a redundancy to partially emphasize the complex biodiversity in tropical regions. The imbalance of number of PFTs among biomes is inherited from a region-specific version of SEIB-DGVM, and will be updated in the future versions of sSEIB.

**Table 1. A part of model parameters for 13 plant
functional types (PFTs) represented in sSEIB**

Climatic optimum and leaf traits	potential assimilation rate [μmolC/m^2/s]	Specific leaf area [gC/m^2]	Leaf nitrogen fraction	Probability of establishment	Minimum GDD [°C • day]
	A_{max}	SLA	N_f	$P_{establish}$	GDD_{min}
Tropical broadleaf evergreen type 1	14.1	0.01	0.016	0.01	2000
Tropical broadleaf evergreen type 2	14.6	0.01	0.016	0.0015	2000
Tropical broadleaf evergreen type 3	33.0	0.01	0.016	0.03	2000
Tropical broadleaf evergreen type 4	22.1	0.01	0.016	0.0015	2000
Tropical broadleaf raingreen	14.1	0.013	0.022	0.015	2000
Temperate needleleaf evergreen	9.0	0.004	0.012	0.04	900
Temperate broadleaf evergreen	9.0	0.007	0.012	0.04	1200
Temperate broadleaf summergreen	12.0	0.015	0.022	0.013	1200
Boreal needleleaf evergreen	9.3	0.004	0.012	0.005	600
Boreal needleleaf summergreen	13.0	0.014	0.016	0.015	350
Boreal broadleaf summergreen	9.0	0.016	0.025	0.02	350
Temperate herbaceous	8.0	0.007	0.027	-	-
Tropical herbaceous	13.0	0.007	0.018	-	-

Ecophysiology

The rate of gross photosynthesis per unit leaf area is determined by the potential assimilation rate of each PFT multiplied by environmental constraints (i.e., temperature, soil moisture, and CO_2 concentration). The C gain of individual plants is a function of leaf area, which is estimated by leaf biomass using specific leaf area (SLA). Respiration of plant tissues is a function of tissue biomass and tissue type. Since foliage respiration rates have a wide range, they are also function of the PFT-specific leaf nitrogen content (N_f: Table 1).

Table 2. Soil decomposition parameters

Parameter	Symbol	Unit	value
SOC turnover, fast	k_{f0}	yr-1	0.3
SOC turnover, slow	k_{s0}	yr-1	0.02

Population and Community Dynamics

In each cell of global grid system, tree dynamics are simulated in a representative stand of 30 m ×30 m and extrapolated over the grid cell. Unlike the state-of-the-art, three-dimensional forest light structure of SEIB-DGVM, sSEIB uses a simplified light attenuation function based on the Beer-Lambert law, in order to facilitate the simulation in a computationally demanding earth system model. However, sSEIB is still a height-structured forest model, and individual trees grow and compete in one dimensional, vertical light gradients of the representative stand. Seedling establishment is reproduced by PFT-specific probability of establishment ($P_{establish}$: Table 1). The survival of seedling is constrained by the climate, using the requirement of minimum growth degree-day (GDD_{min}: Table 1). sSEIB is still spatially explicit in three dimensions, regarding canopy occupation of a space. Individual tree canopies of definite sizes also compete for space, and a sub-canopy tree may not have an access to the top layer unless a gap is created.

Soil Organic Carbon (SOC)

SOC is the key variable of the global C cycle due to its large stock and potential sensitivity to environmental conditions (Dixon et al. 1994). To effectively simulate SOC, the multi-pool approach is used in sSEIB. There are fast- and slow-decomposing SOC with different potential turnover rates k_{f0} and k_{s0}, respectively, adopted from Roth-C (Coleman and Jenkinson 1996; Table 2). The fast SOC responds quickly to fast-timescale environmental changes such as interannual variabilities. The overall response of the slow SOC pool will be determined by long-term trends of environmental conditions such as soil temperature and moisture. We assume that this multi-pool formulation has an ability to realistically show the transient behavior of SOC. The potential SOC turnover rates are then calculated by multiplying soil temperature (TD) and moisture (MD) modifiers (Figure 2):

$$TD = \exp\left[308.56 \cdot \left(\frac{1}{66.02} - \frac{1}{T_{soil} + 46.02}\right)\right]$$

$$MD = 0.25 + 0.75 \cdot W_{soil}$$

where T_{soil} is the mean soil temperature [°C] of the first layer (0-0.05 m) and W_{soil} is the degree of saturation. TD is standardized to $T_{soil} = 20°C$ where $TD = 1$. The realized decomposition rates are:

$$k_f = k_{f0} \cdot TD \cdot MD$$

$$k_s = k_{s0} \cdot TD \cdot MD$$

where k_f and k_s are realized decomposition rates of fast- and slow-decomposing SOC pools, respectively. The input to the fast SOC pool is freshly dead vegetation biomass, originated from both litterfall and death of individual trees. The fast SOC pool is decomposed by a microbial efficiency f_m of 0.7 (Parton et al. 1987; Ise and Moorcroft 2006) by which the decomposed SOC released to the atmosphere as CO_2. The remaining SOC is transferred into the slow SOC pool. Thus, C balance of fast and slow SOC, C_f and C_s, respectively:

$$\frac{dC_f}{dt} = L - k_f \cdot C_f$$

$$\frac{dC_s}{dt} = (1 - f_m) \cdot k_f \cdot C_f - k_s \cdot C_s$$

where L is C input from vegetation.

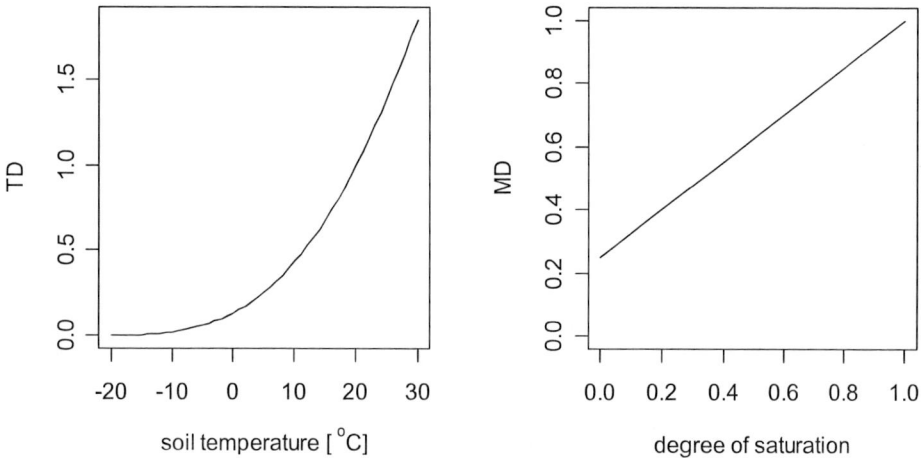

Figure 2. Temperature and moisture dependency curves of SOC (soil organic carbon) decomposition.

Land Use Change

Because of its extent and intensity (Vitousek et al. 1997), anthropogenic land use will induce large land-atmosphere feedbacks (Dale 1997; Hurtt et al. 2006). For example, an agricultural conversion of temperate forest significantly increase surface albedo and decrease evapotranspiration, and the local energy balance such as Bowen ratio is largely altered. Moreover, the human appropriation of land modifies C storage in both above- and belowground. Removal of woody species reduces C stock in living biomass. The amount of SOC is usually reduced due to a reduction in litter input and intensive soil management such as tillage.

Land use change (LUC) is also explicitly treated by sSEIB according to the historical and projected datasets by Hurtt et al. (2006). The historical and projected land surfaces are categorized into primary forest, secondary forest, cropland, and pasture, and fractions of these land use types are provided for each year in 1700-2100, in 0.5° resolution. Since sSEIB is an individual-based model, it simulates tree population dynamics arising from the anthropogenic land use mechanistically. For example, the development of secondary forests after logging and agricultural abandonment is explicitly simulated by sSEIB, including individual tree growth patterns and successional dynamics. The model also has a capacity to represent size- and/or PFT-specific selective cutting.

Coupling to a General Circulation Model (GCM)

To explicitly reproduce the two-way feedback between the terrestrial ecosystem and the climate (Moorcroft 2003), sSEIB is now coupled to CCSR-FRCGC GCM (Kawamiya et al. 2005). The meteorological forcings (air temperature, soil temperature, precipitation, wind speed, vapor pressure deficit, etc.) are calculated by the GCM and passed to sSEIB, which simulates the dynamics of terrestrial ecosystem based on the meteorological inputs. Then, vegetation statuses such as LAI are passed to CCSR-FRCGC GCM through MATSIRO to simulate land-surface biophysics (e.g., balances of surface energy and water). Moreover, the terrestrial C balance arisen from sSEIB affects atmospheric concentrations of CO_2 that determines radiative forcing of the GCM. We adopted t42 grid resolution (64 latitudinal and 128 longitudinal grids to cover the earth surface) to represent the global earth surface.

The land-surface model MATSIRO acts as a biophysical coupler between the atmospheric and ecosystem models. MATSIRO simulates land-surface physical processes such as radiative balance, energy budget, heat transfer, and water cycling and calculates variables such as soil temperature and moisture based on atmospheric forcing from the GCM. Biophysical conditions and processes such as albedo and surface hydrological and energy balances are simulated based on vegetation predicted by sSEIB, as these variables can be represented by functions of vegetation type and quantity (i.e., LAI). Gaps in time steps are filled by the coupler module, which controls the variable traffic with different calculation time steps of sub-daily atmospheric and land-surface processes and daily/monthly/annual ecosystem processes.

Study Design

Compared to other TBMs that are coupled to GCMs, sSEIB explicitly simulates plant population dynamics. Therefore, the transient responses such as time lags in biosphere responses to climate change can be realistically reproduced, based on processes of individual plant population and community ecology—seed dispersal and establishment, growth, survival, competition, and mortality. This characteristic allows us to make various novel and meaningful simulations regarding climate change. For example, tracking the boreal-tundra ecotone movement will have strong implications to global radiative balance. The mechanistically dynamic vegetation can be used to reproduce paleo-biogeography of the Holocene, including the vegetation-atmosphere interactions.

One of the advantages of simulation modeling is to make a set of experimental simulations to quantitatively attribute ecological and atmospheric changes to different mechanisms. Here we have several plans. (1) By systematically changing rates of seed dispersal and seedling establishment, we will quantify the effect of vegetation shift onto the climate change. (2) In sSEIB, mortality is assigned individually based on PFT and size, in addition to C balance. Experimental modification in interannual variability in climate may cause biome- and PFT-specific changes in mortality, and the resultant atmospheric feedback can be altered. (3) The past, current, and future human land use has strong impacts on ecosystem function. We compare simulations without land use (potential vegetation) and with land use (actual vegetation) and quantify the changes in biophysics and biogeochemistry. The GCM-coupled model system (Figure 1) has a capacity to switch on and off biophysical and biogeochemical coupling individually, allowing us to separate factors.

Current Results

We made two sets of simulations: (1) decoupled and (2) coupled simulations with GCM. First, in the decoupled simulation, to observe the model's fundamental behavior under a constant climate, we forced the stand-alone sSEIB with an observed climate of 1961-1990 (New et al. 2000) repeatedly for 2000 years. After the establishment of equilibrium conditions of vegetation and SOC, we disturbed the model by an instantaneous increase in temperature by 3°C for 300 years, to study model's sensitivity to climate change. This decoupled simulation only showed climate-to-ecosystem forcing (no two-way feedback) as the climate was from the external data of 1961-1990. Then we coupled sSEIB with CCSR-FRCGC GCM to study the effects of two-way feedback (both climate-to-ecosystem and ecosystem-to-climate) for 100 years, under the pre-industrial concentration of atmospheric CO_2.

Decoupled Simulation

The decoupled simulation of 2000 years from bareground gave equilibria of global C pools (Figure 3). The biomass in vegetation reached the equilibrium relatively quickly, around year 100. Soon after the stabilization of vegetation, the fast SOC pool also reached the equilibrium as the litter input from vegetation became constant. It took more than 1000 years for slow SOC to be stabilized, however, since the turnover time of this SOC pool is much longer than the fast SOC (Table 2). Due to the sudden warming at year 2000, the global biomass decreased by 30%. Because of the individual-based nature of sSEIB, the model reproduced effects of the heat shock—aggravated plant C balance, increase in water loss, replacement of old-growth forests with young stands of new PFTs—that all resulted in a large die-off of woody biomass. After the temperature shock (years 2050-2300), the system showed a partial recovery from the disturbance, toward the new equilibrium under the warmer climate. As in the most terrestrial ecosystem models (IPCC 2007), sSEIB showed an overall decrease in global biomass due to warming. However, the transient behavior of our model—

large die-off and partial recovery—was the novel result from an individual-based model. Of course, the realistic warming will be gradual, and the magnitude of the die-off may be smaller than this experimental simulation.

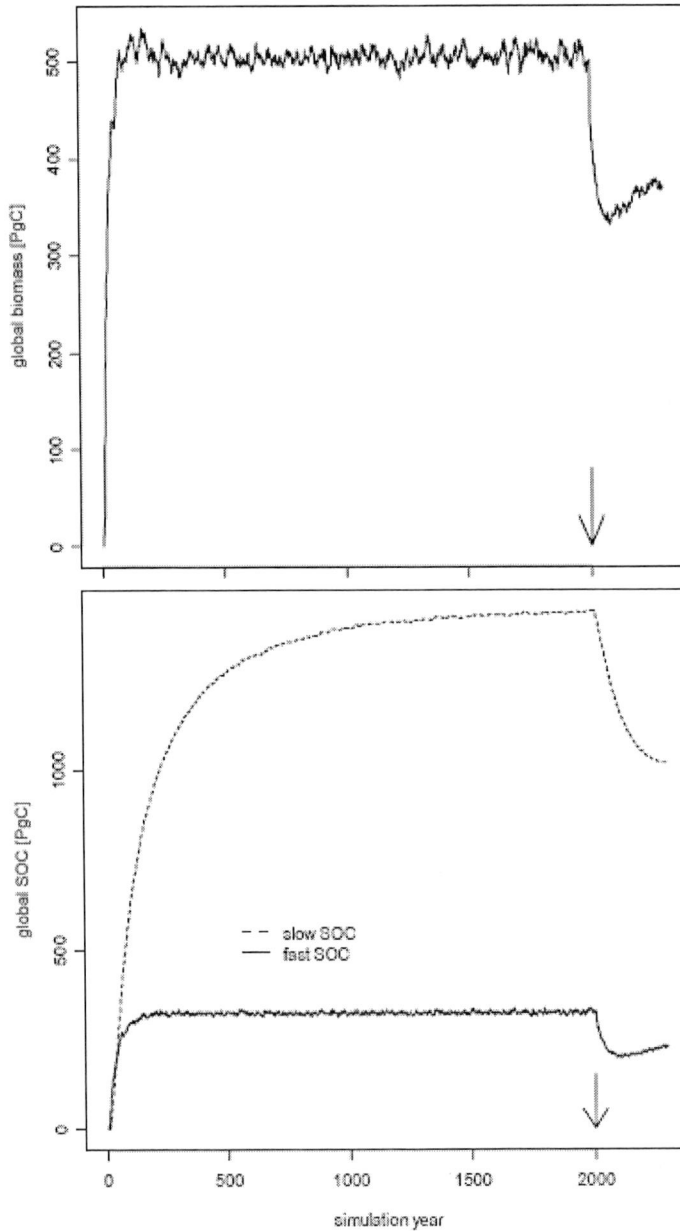

Figure 3. Trajectories of C pools, decoupled simulation of 2300 years. Arrows at simulation year 2000 represent the timing of experimental warming of 3°C.

Regional patterns of the decoupled simulations are shown in global maps (Figure 4). The maps at year 2000 represent the equilibrium conditions under the current climate (Figs. 4a and 4d). Note that the vegetation biomass is concentrated in biomes that are dominated by forests. SOC accumulation was strong in boreal regions, as in the observation (Ise and Moorcroft 2006). Due to the stochastic nature of an individual-based model, outputs from sSEIB are often patchy especially in closed-canopy forests. In general, the experimental warming reduced the plant biomass (Figure 4b and 4c), but there was a significant increase in biomass in tundra. Due to the increase in soil temperature, the microbial activity of SOC decomposition was enhanced, and a reduction in SOC was observed (Figure 4e and 4f).

The decrease in litter input due to the biomass reduction is another cause of the global trend in SOC reduction. However, due to the northward movement of forest-tundra ecotone, the southern edge of tundra became a significant C sink.

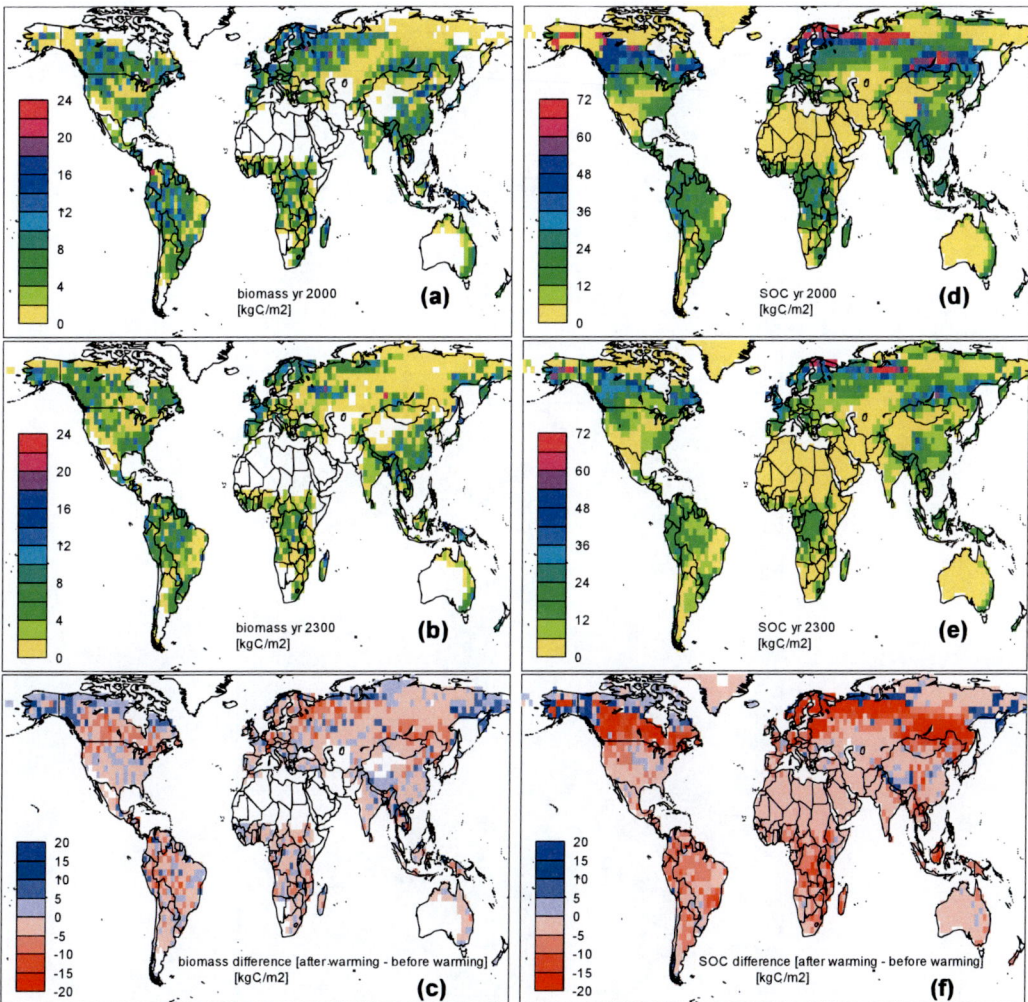

Figure 4. Global distributions of C stocks under the current climate (a and d) and under the experimental warming of 3°C (b and e). Gains and losses of C due to the warming is also shown in difference maps (c and f), with the shades of blue and red represent the magnitudes of C gain and loss, respectively.

Coupled Simulation

To verify the model's performance, here we show the results from a GCM-coupled simulation under the pre-industrial CO_2 concentration, prescribed sea surface temperature, and sea ice extent, for 100 years (Figure 5). In this simulation, the land surface (vegetation) and the atmosphere are coupled by the two-way feedback. Within the simulation year of 100 years, the ecosystem variables (vegetation distribution, C pool in biomass and SOC, and C flux of net primary production and ecosystem respiration) was brought into a dynamic equilibrium under the constant forcing, and the corresponding atmospheric condition was evolved.

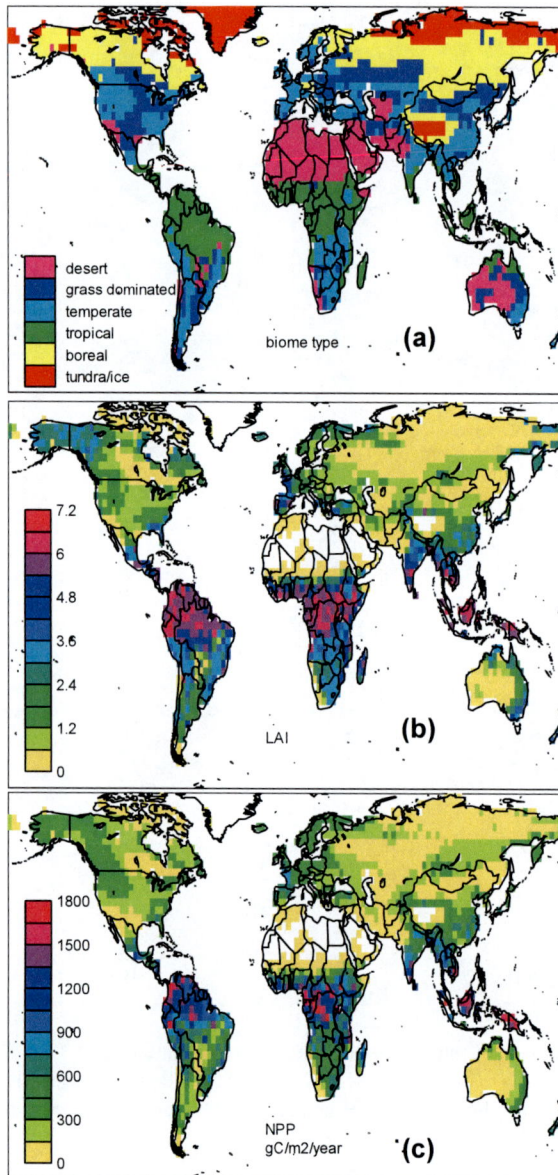

Figure 5. Output from a coupled simulation. Global distributions of (a) biome, (b) LAI, and (c) NPP.

The global distribution of biome for the pre-industrial condition, or potential natural vegetation, broadly matched the observation (Figure 5a). LAI (Figure 5b) and NPP (net primary productivity: Figure 5c) had general latitudinal gradients, and were lowered in arid regions (desert and herb-dominated areas). In a coupled model, it is especially important to reproduce regional patterns of NPP as the spatiotemporal patterns in NPP controls the C flux from atmosphere to biosphere and the resultant radiative forcing from CO_2 concentrations. In a coupled model, LAI is another key variable, and it is directly transferred from sSEIB to MATSIRO. In sSEIB, LAI translates gross photosynthetic rates per unit leaf area into those per unit land area. Foliage respiration rates are also proportional to LAI, and then the overall plant C balance (i.e., NPP) is the key element in biogeochemistry. LAI in MATSIRO affects land surface and atmospheric conditions especially in a regional scale, through sensible and latent heat fluxes from the land surface. In addition, it is helpful to confirm NPP and LAI with observational patterns, as these representative variables are overall outcomes calculated from many underlying ecological processes.

Future Refinements

To increase sSEIB's predictability and model usability, we plan several refinements. First, subgrid-scale heterogeneity within large GCM grid cells should explicitly be considered (Ise and Sato 2008). Ecosystem structure, composition, and function are often radically diverse even within the same climatic regimes, if the local environmental conditions—soil type, geomorphology, slope, and aspect, for example—are different. These unique vegetation statuses should be treated as semi-independent patches within a large GCM grid cell, and the coupling to the atmosphere should be from weighted averages of outcomes from all land-surface patches. However, this multi-patch approach inevitably increases the computational burden in the earth system model. Another potential issue is robustness in model predictions. Since sSEIB estimates ecosystem conditions of the entire GCM grid cell (a rectangle of a few degrees) from a small (30 m × 30 m) representative stand. Therefore, if stochastic disturbance on the stand creates a large gap in the representative stand, ecological variable of the entire grid cell can be radically altered. Although it is theoretically possible to repeat the stochastic simulation for many times to obtain an average, this method is also computationally demanding. Thus, one of the possible solution is the size- and age-structured (SAS) simulation of forest dynamics applied in ED (Moorcroft et al. 2001) to efficiently simulate the global terrestrial biosphere (Purves and Pacala 2008). By applying the SAS concept, sSEIB can realistically accommodate the multi-patch representation of subgrid-scale heterogeneity with a reduction in computational time. Moreover, since the SAS concept emulates the average behavior of numbers of stochastic simulation runs, the results tend to be concentrated around the mean and normally distributed (the central limit theorem).

Due to its vast storage size, climate responses of SOC are expected to have a significant feedback onto the atmosphere through biogeochemical cycling (Ise and Moorcroft 2006; Ise et al. 2008). In spite of the importance, submodels of SOC accumulation and decomposition in most current TBMs are not structured in detail. In our current formulation, the decomposition rate of SOC is only dependent on temperature and moisture of the first soil layer (0-0.05 m). However, our model system (Figure 1) has a detailed land-surface submodel MATSIRO that simulates depth-structured soil physics. By coupling SOC biogeochemical model with the depth-structured soil model, the different climate responses of SOC in vertical layers can be simulated. For example, in permafrost, SOC decomposition rates of the active layer and perennially frozen layer are different in many orders of magnitude. The soil temperature profile reproduced by MATSIRO can simulate the SOC gradient under the climate change in a transient manner.

Fire has a strong impact on the land-atmosphere feedback (Randerson et al. 2006). Fire instantaneously combusts large amount of live biomass and organic litter and emits CO_2 to the atmosphere (biogeochemical effects). Fire also changes surface energy and water balance such as albedo and evapotranspiration (biophysical effects). However, predicting the future fire regime has been extremely difficult, due to the highly stochastic nature of fire ignition and propagation. The physical representation of land surface such as litter amount, temperature, and moisture from the earth system model is expected to greatly improve the projection of the future fire patterns. Moreover, human management of fire such as fire suppression and prescribed fire should explicitly be accounted for by adopting an appropriate land use dataset.

Nutrient cycling is another important factor on vegetation modeling under the global change (Bonan 2008B; Sokolov et al. 2008). For example, availability of nutrient often determines the degree of CO_2 fertilization (Oren et al. 2001). The differential vegetation responses due to nutritional status should be incorporated in the above multi-patch simulation concept. Nutrient availability is also coupled to the SOC decomposition submodel to increase the system integrity.

In conclusion, local ecological dynamics are very important factors that determine the land-atmosphere interaction, and the individual-based approach of forest modeling is a useful tool to realistically reproduce the ecological responses and forcings. As we show in the decoupled simulation (Figures 3 and 4), the stand-level effects of disturbance and recovery from it will have a large impact on the global earth system. The updates on sSEIB will improve the predictability of the coupled climate simulations.

ACKNOWLEDGMENTS

This study was funded by Innovative Program of Climate Change Projection for the 21st Century of the Ministry of Education, Culture, Sports, Science and Technology (MEXT).

REFERENCES

Albani, M., Medvigy, D., Hurtt, G. C., and Moorcroft, P. R. (2006). The contributions of land-use change, CO2 fertilization, and climate variability to the Eastern US carbon sink. *Global Change Biology*, 12, 2370-2390.
Betts, R. A., Cox, P. M., Lee, S. E., and Woodward, F. I. (1997). Contrasting physiological and structural vegetation feedbacks in climate change simulations. *Nature*, 387, 796-799.
Bonan, G. B., Levis, S., Sitch, S., Vertenstein, M., and Oleson, K. W. (2003). A dynamic global vegetation model for use with climate models: concepts and description of simulated vegetation dynamics. *Global Change Biology*, 9, 1543-1566.

Bonan, G. B. (2008A). Forests and climate change: Forcings, feedbacks, and the climate benefits of forests. *Science*, 320, 1444-1449.

Bonan, G. (2008B). Carbon cycle: Fertilizing change. *Nature Geoscience,* 1, 645-646.

Botkin, D. B., Wallis, J. R., and Janak, J. F. (1972). Some Ecological Consequences of a Computer Model of Forest Growth. *Journal of Ecology*, 60, 849-and.

Charney, J. G. (1975). Dynamics of Deserts and Drought in Sahel. *Quarterly Journal of the Royal Meteorological Society,* 101, 193-202.

Coleman, K., and Jenkinson, D. S. (1996). A model for the turnover of carbon in soil. In D. S. Powlson and P. Smith and J. U. Smith (Eds.), Evaluation of soil organic matter models using existing, long-term datasets. NATO ASI Series I (Vol. 38, pp. 237-246). New York: Springer.

Cox, P. (2001). Description on the Triffid Dynamic Global Vegetation Model, Technical Report 24. Devon, UK: Hadley Centre Met Office.

Dale, V. H. (1997). The relationship between land-use change and climate change. *Ecological Applications*, 7, 753-769.

Delworth, T. L., Broccoli, A. J., Rosati, A., et al. (2006). GFDL's CM2 global coupled climate models. Part I: Formulation and simulation characteristics. *Journal of Climate*, 19, 643-674.

Dixon, R. K., Brown, S., Houghton, R. A., et al. (1994). Carbon Pools and Flux of Global Forest Ecosystems. *Science,* 263, 185-190.

Foley, J. A., Costa, M. H., Delire, C., Ramankutty, N., and Snyder, P. (2003). Green surprise? How terrestrial ecosystems could affect earth's climate. *Frontiers in Ecology and the Environment,* 1, 38-44.

Hurtt, G. C., Frolking, S., Fearon, M. G., et al. (2006). The underpinnings of land-use history: three centuries of global gridded land-use transitions, wood-harvest activity, and resulting secondary lands. *Global Change Biology,* 12, 1208-1229.

IPCC. (2007). Climate change 2007: the physical science basis. In S. Solomon et al. (Eds.), Contribution of Working Group I to the Fourth Assessment Report of the Intergovernmental Panel on Climate Change. Cambridge, UK and New York, NY, USA: Cambridge University Press.

Ise, T., and Moorcroft, P. R. (2006). The global-scale temperature and moisture dependencies of soil organic carbon decomposition: an analysis using a mechanistic decomposition model. *Biogeochemistry*, 80, 217-231.

Ise, T., and Moorcroft, P. R. (2008). Quantifying local factors in medium-frequency trends of tree ring records: Case study in Canadian boreal forests. Forest Ecology and Management, 256, 99-105.

Ise, T., Dunn, A. L., Wofsy, S. C., and Moorcroft, P. R. (2008). High sensitivity of peat decomposition to climate change through water-table feedback. *Nature Geoscience*, doi:10.1038/ngeo1331.

Ise, T., and Sato, H. (2008). Representing subgrid-scale edaphic heterogeneity in a large-scale ecosystem model: A case study in the circumpolar boreal regions. *Geophysical Research Letters*, 35, L20407, doi:10.1029/2008GL035701.

Ito, A., and Oikawa, T. (2002). A simulation model of the carbon cycle in land ecosystems (Sim-CYCLE): A description based on dry-matter production theory and plot-scale validation. *Ecological Modelling*, 151, 143-176.

Kawamiya, M., Yoshikawa, C., Kato, T., et al. (2005). Development of an integrated earth system model on the Earth Simulator. Journal of the Earth Simulator, 4, 18-30.

Levis, S., Foley, J. A., and Pollard, D. (1999). Potential high-latitude vegetation feedbacks on CO2-induced climate change. *Geophysical Research Letters*, 26, 747-750.

Luo, Y. Q. (2007). Terrestrial carbon-cycle feedback to climate warming. *Annual Review of Ecology Evolution and Systematics*, 38, 683-712.

Moorcroft, P. R., Hurtt, G. C., and Pacala, S. W. (2001). A method for scaling vegetation dynamics: The ecosystem demography model (ED). *Ecological Monographs*, 71, 557-585.

Moorcroft, P. R. (2003). Recent advances in ecosystem-atmosphere interactions: an ecological perspective. *Proceedings of the Royal Society B-Biological Sciences*, 270, 1215-1227.

Moorcroft, P. R. (2006). How close are we to a predictive science of the biosphere? *Trends in Ecology and Evolution*, 21, 400-407.

New, M., Hulme, M., and Jones, P. D. (2000). Global 30-year mean monthly climatology, 1961-1990 data set (Available online [http://www.daac.ornl.gov] from Oak Ridge National Laboratory Distributed Active Archive Center, Oak Ridge, TN).

Oren, R., Ellsworth, D. S., Johnsen, K. H., et al. (2001). Soil fertility limits carbon sequestration by forest ecosystems in a CO2-enriched atmosphere. *Nature,* 411, 469-472.

Pacala, S. W., Canham, C. D., and Silander, J. A. (1993). Forest Models Defined by Field-Measurements .1. The Design of a Northeastern Forest Simulator. Canadian Journal of Forest Research-Revue Canadienne De Recherche Forestiere, 23, 1980-1988.

Parton, W. J., Schimel, D. S., Cole, C. V., and Ojima, D. S. (1987). Analysis of factors controlling soil organic-matter levels in Great-Plains grasslands. *Soil Science Society of America Journal*, 51, 1173-1179.

Purves, D., and Pacala, S. (2008). Predictive models of forest dynamics. Science, 320, 1452-1453.

Randerson, J. T., Liu, H., Flanner, M. G., et al. (2006). The impact of boreal forest fire on climate warming. *Science,* 314, 1130-1132.

Sato, H., Itoh, A., and Kohyama, T. (2007). SEIB-DGVM: A new dynamic global vegetation model using a spatially explicit individual-based approach. *Ecological Modelling*, 200, 279-307.

Shugart, H. H. (1984). A theory of forest dynamics. New York: Springer-Verlag.

Sitch, S., Smith, B., Prentice, I. C., et al. (2003). Evaluation of ecosystem dynamics, plant geography and terrestrial carbon cycling in the LPJ dynamic global vegetation model. *Global Change Biology,* 9, 161-185.

Sokolov, A. P., Kicklighter, D. W., Melillo, J. M., et al. (2008). Consequences of Considering Carbon-Nitrogen Interactions on the Feedbacks between Climate and the Terrestrial Carbon Cycle. *Journal of Climate,* 21, 3776-3796.

Takata, K., Emori, S., and Watanabe, T. (2003). Development of the minimal advanced treatments of surface interaction and runoff. Global and Planetary Change, 38, 209-222.

Vitousek, P. M., Mooney, H. A., Lubchenco, J., and Melillo, J. M. (1997). Human domination of Earth's ecosystems. *Science,* 277, 494-499.

Whittaker, R. H. (1975). Communities and ecosystems (2 ed.). New York: Macmillan.

In: Forest Canopies: Forest Production, Ecosystem…   ISBN 978-1-60741-457-5
Editor: J. D. Creighton and P. J. Roney   © 2009 Nova Science Publishers, Inc.

Chapter 8

ATMOSPHERIC DEPOSITION AND ITS LEAF SURFACE INTERACTIONS IN JAPANESE CEDAR FORESTS

Hiroyuki Sase[1] and Takejiro Takamatsu[2]

[1.] Ecological Impact Research Department, Acid Deposition and Oxidant Research Center, 1182 Sowa, Nishi-ku, Niigata 950-2144, Japan;
[2.] Center for Water Environment Studies, Ibaraki University, 1375 Ohu, Itako, Ibaraki 311-2402, Japan

ABSTRACT

Forest canopy may be an important interface between atmosphere and forest ecosystems. Properties of leaf surface as a major part of the forest canopy were studied mainly in Japanese cedar (*Cryptomeria japonica*) forests. The amount of epicuticular wax increased under the effect of water stress (on high branches and at locations with low rain factors), exposure to the noxious gases (such as volcanic acidic gases), and strong UV radiation at high altitude, while the C content of wax decreased and the O content increased, except in case of the altitude. In the Kanto Plain around Tokyo, where Japanese cedar is declining, the wax eroded more rapidly (approximately 1.5 times faster) than that of healthy trees in mountainous areas, although the amount of wax in current-year leaves was almost equivalent in both areas. Amounts of anthropogenic elements such as antimony (Sb) in particulate matters deposited on the leaf surface were greater (10 times greater in case of Sb) at the severe decline area than at the healthy area. Atmospheric deposition including the particulate matters may be a possible cause of the wax degradation. Fractions of unhealthy stomata (disability in closing) correlated with the amounts of particulate Sb on the leaves. In fact, stomata clogged with particles were observed by optical or scanning electron microscopes. The cuticular transpiration rate in 1-year leaves was higher in the plain area (0.92% h^{-1} as decreasing rate of leaf water content) than in the mountainous area (0.60% h^{-1}). This water loss, resulting from a degraded wax layer and partial malfunctioning of stomata due to deposited particulate matters, may be a significant factor causing the decline of Japanese cedar. Recent dry atmospheric conditions in the plain area may accelerate the water stress of trees. Anthropogenic elements in the particulate matters on the leaf can be utilized as indicators

---

[1] Corresponding author (e-mail: sase@adorc.gr.jp).

of air pollution, too. The amounts of Sb correlated with NO_X concentration and population density in each sampling area. Moreover, epicuticular wax properties and particulate matters on the leaf may affect leaf surface wettability. Increase in leaf wettability may accelerate ion exchange on the leaf surface, resulting in increase of leaching of K^+ and uptake/consumption of N compounds on the forest canopy. The canopy interactions should be considered for discussion of elemental flows in forest ecosystems. Small changes in leaf surface properties on the forest canopy may affect plant physiology and biogeochemical cycles of elements in forest ecosystems.

1. INTRODUCTION

Forest canopy may be an important interface between atmosphere and forest ecosystems. In particular, the leaf surface covered by the cuticle layer, as a major part of the forest canopy, plays an important role in defense against water loss, ion penetration and leaching, and fungal infection (Trunen and Huttunen, 1990). Therefore, changes in leaf surface properties may cause serious effects on the physiological functions of trees sometimes (Sase et al., 1998b) or affect ion fluxes on the canopy (Sase et al., 2008).

A significant decline (dieback and/or defoliation) of Japanese cedar (*Cryptomeria japonica*) has been observed in the Kanto Plain around Tokyo since the 1960s. As possible factors causing the tree decline, the direct effects of oxidants (Takahashi et al., 1987), soil acidification (Nashimoto et al., 1993), and the effect of water deficiency (Matsumoto et al., 1992; Sakata 1996) had been suggested. Air pollutants could affect the cuticle layer, especially the epicuticular wax, directly or indirectly (Trunen and Huttunen, 1990; Percy et al., 1994). The degradation of wax may cause serious water stress to trees (Sase et al., 1998a).

The surface conditions of the leaves, such as leaf wettability, may also regulate the ion exchange reactions occurring on the canopy. Throughfall (TF) and stemflow (SF) are useful parameters for estimating the total amount of ion deposition in forested areas, as well as the dry deposition of certain constituents (Lindberg and Lovett, 1992). However, the chemical constituents of TF and SF vary depending on canopy interactions, such as leaching, uptake, and/or consumption of ions on the tree surface (Lovett et al., 1985; Shibata and Sakuma, 1996). The mechanisms of canopy interactions are important for evaluating the amount of total and dry deposition in forest ecosystems.

In our studies, changes on leaf surface properties were examined to elucidate significant factors causing the decline of *C. japonica* in the Kanto Plain. Anthropogenic elements in particulate matters on the leaf surface were also analyzed as a possible indicator of air pollution. Moreover, surface properties of individual leaves were discussed as possible factors that regulate TF and SF chemistry in a *C. japonica* forest in Niigata Prefecture, which faces the Sea of Japan and is affected by seasonal winds from the west in winter.

2. SUMMARY OF THE METHODS

Cryptomeria japonica leaves were collected from the areas shown in Figure 1. The leaves were collected with branches according to the method of Sase et al. (1998a). Current-year (0-y) and one-year-old (1-y) leaves were collected for analyses. For comparison with *C.*

japonica, leaves/needles of other conifers, such as *Cedrus deodara*, and *Thujopsis dolabrata* var. *hondai*, were occasionally collected.

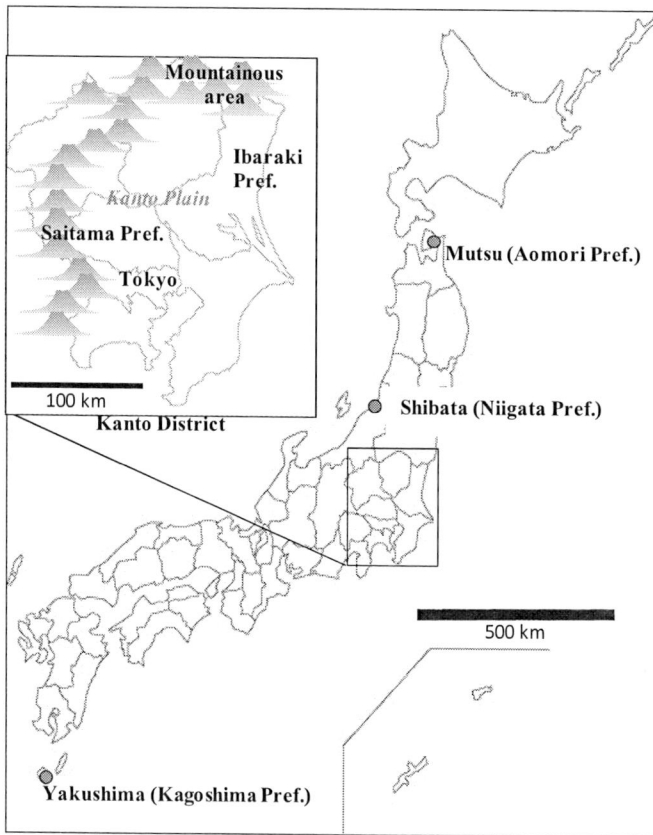

Figure 1. Sampling locations of conifers used in this study. *Cryptomeria japonica* is declining severely in Saitama Prefecture. Volcanic fumaroles, which emit acidic gases, distribute at Osorezan in Mutsu. Yakushima is a small island with limited human activities except for existence of a small-scale electrochemical plant. Catchment experiment to study the canopy interactions was conducted in Shibata.

Epicuticular wax (EW) was extracted by shaking 5-g samples of the leaves for 15 s in 20 mL of $CHCl_3$, and then measured gravimetrically once the $CHCl_3$ evaporated (Sase et al., 1998a). The amount of wax was expressed in units of mg g^{-1} fresh leaves or dry leaves (FL or DL; Sase et al., 1998a).

The C and O contents of the wax (Sase et al., 1998a; 1998b), contact angles (CA) of water droplets on the leaf surface (Takamatsu et al., 2001a; 2001b; Sase et al., 2008), elemental composition of particulate matters deposited on the leaves (Takamatsu et al., 2000; 2001a), leaching rate of K^+ from the leaf surface (Sase et al., 2008), and fluxes of ions from rainfall outside the canopy (RF), throughfall (TF), and stemflow (SF) (Sase et al., 2008) were measured for the respective studies. SEM observation (Sase et al., 1998b; 2008; Takamatsu et al., 2001a) and detection of unhealthy stomata (Takamatsu et al., 2001a) were also carried out.

3. EPICUTICULAR WAX PROPERTIES AND ENVIRONMENTAL FACTORS

The properties of the epicuticular wax from *C. japonica* leaves varied because of natural environmental factors such as altitude and exposure to noxious gases (e.g. volcanic acidic gases), as well as branch height and foliar age within the tree. However, changes in C and O contents of the wax showed different patterns from those in the wax amounts occasionally, depending on effective factors.

3.1. Leaf Age

Figure 2 shows changes in the amounts, and C and O contents of epicuticular wax in *C. japonica* and *C. deodara* during the growing phase. The amounts of wax from the 0-y leaves increased significantly during the initial two months of growing season, from May to July. These reached peaks in summer, and then decreased gradually (Sase et al., 1998a). Although the water content of 0-y leaves also decreased during the growing phase, similar profiles were maintained in the seasonal changes in the amount of wax on a dry-mass basis (Sase et al., 1998a). The C content decreased significantly during the growing phase, while the O contents increased significantly (Figure 2). The C contents reached a minimum in autumn, and then increased gradually (Sase et al., 1998a).

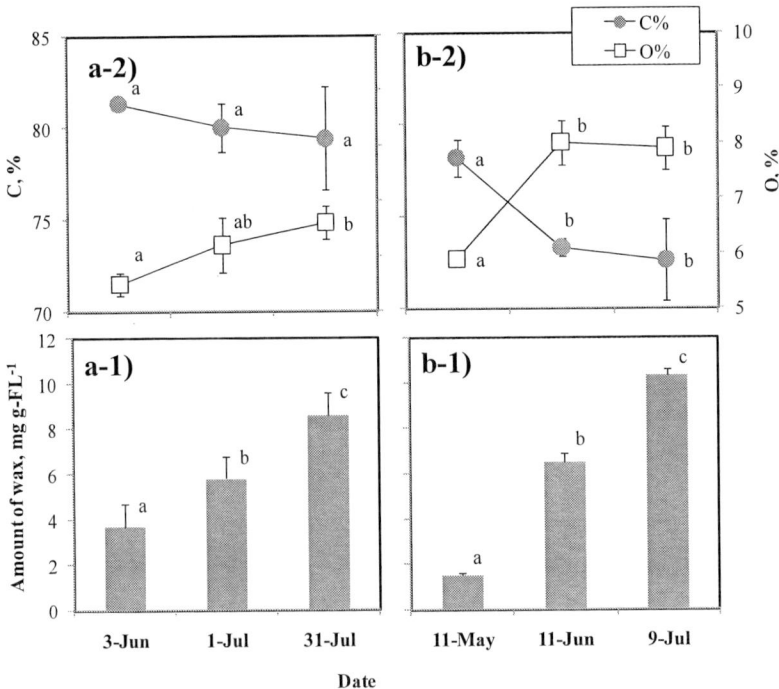

Figure 2. Changes in amount (1) and C and O contents (2) of epicuticular wax in *Cryptomeria japonica* (a) and *Cedrus deodara* (b) during the growing phase in 1992 (redrawn based on Sase et al. (1998a)). Plots show means and standard deviations of triplicate analyses. FL: fresh leaves. Different letters beside the plots indicate significant ($p < 0.05$) differences.

It was suggested that the wax constituents were deposited in an ordered sequence with a tendency for more polar wax classes to be synthesized later during leaf expansion (Percy et al., 1994). The changes in C and O contents during the growing phase may reflect the time dependency of wax biosynthesis. On the other hand, the gradual increase in C content during the latter phase may be due to the selective erosion of specific wax constituents in the natural and anthropogenic process of wax degradation (Sase et al., 1998a).

3.2. Altitude

Changes in properties of epicuticular wax in 1-y leaves of *C. japonica* with increasing altitude on Yakushima Island are shown in Figure 3. The same tendency was also seen on both a dry-mass basis and a leaf surface area basis (Sase et al., 1998a). The amount and C content of wax increased with increasing altitude, while the O content decreased significantly (Figure 3a and b).

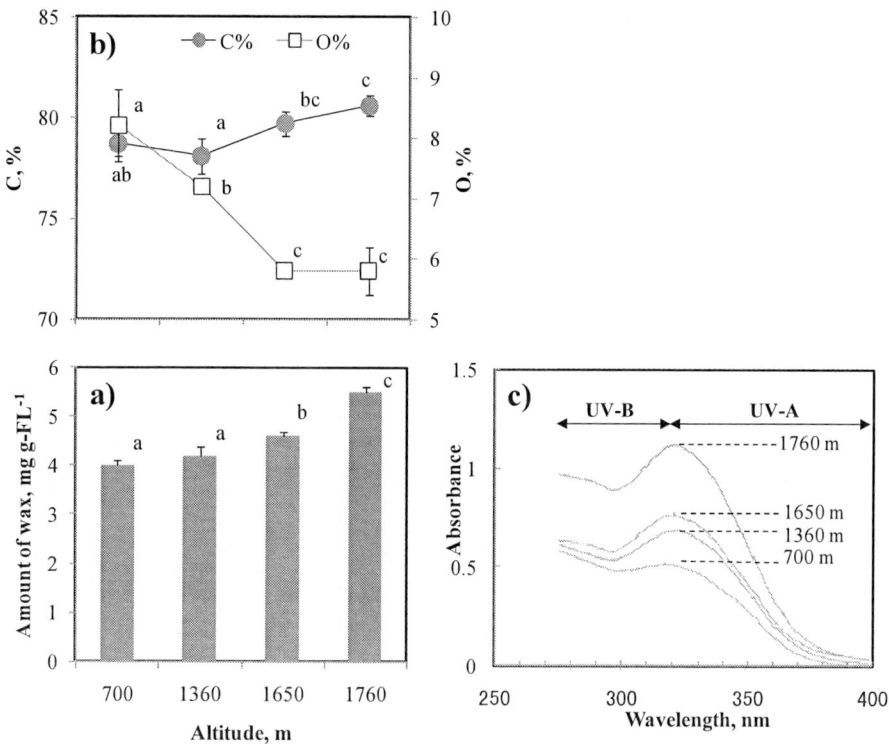

Figure 3. Changes in amount (a), C and O contents (b), and UV absorption spectra (c) of epicuticular wax in *Cryptomeria japonica* with increasing altitude on Yakushima Island (redrawn based on Sase et al. (1998a)). Plots show means and standard deviations of triplicate analyses. FL: fresh leaves. Different letters beside the plots indicate significant (p < 0.05) differences.

An increase in the amount of wax with altitude has been reported also for other species (e.g. Günthardt-Goerg, 1994). It was usually suggested that the increase in the wax amount was related to water deficiency, since epicuticular wax plays an important role in conservation of water in leaves. However, in case of *C. japonica* in Yakushima, not only water condition but also strong UV radiation may affect the wax production (Sase et al, 1998a). The UV absorption spectra of the wax from 1-y leaves at different altitudes are shown in Figure 3c. The wax from trees at higher altitudes absorbed more UV radiation. The UV absorption spectra observed here were similar to those of the unbound free flavonoids present in surface wax (DeLucia et al., 1992). It was suggested that the strong UV radiation at higher altitudes increased the production of either epicuticular wax or flavonoids in *C. japonica* leaves (Sase et al., 1998a).

3.3. Branch Height

Changes in properties of epicuticular wax with increasing branch height in 1-y leaves of *C. japonica* and *Thujopsis dolabrata* are shown in Figure 4. The amount of wax increased with increasing branch height on a fresh-mass basis in both species, while the tendency in *T. dolabrata*-A was not clear. The same tendency was also seen on both a dry-mass basis and a leaf surface area basis, and this phenomenon was most evident in *C. japonica* (Sase et al., 1998a). The C content of epicuticular wax decreased with increasing branch height in both species, while the O content increased.

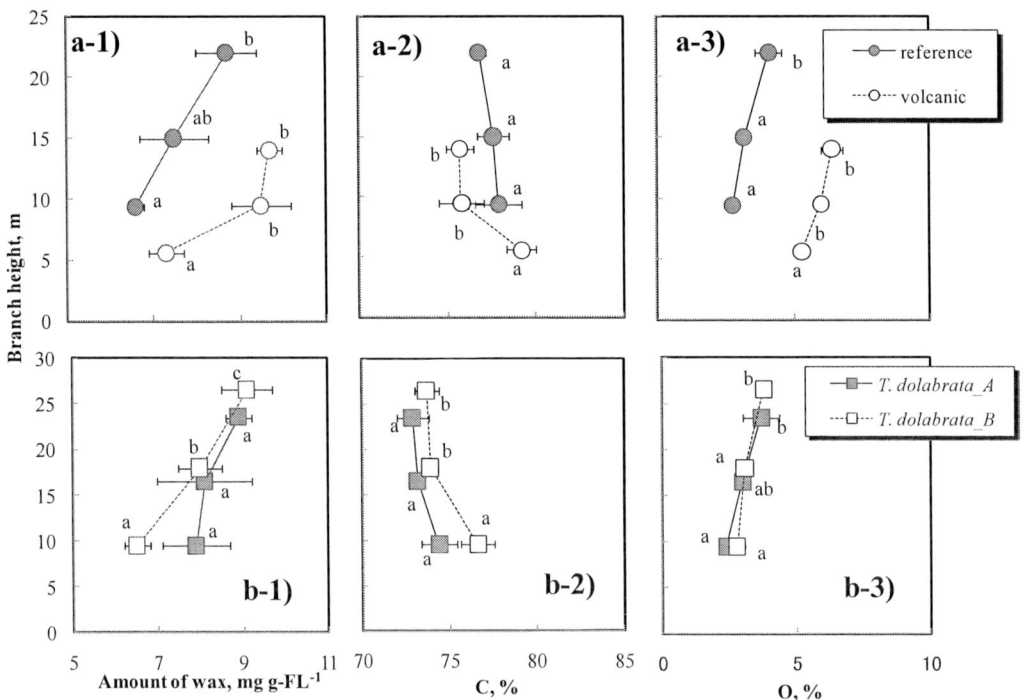

Figure 4. Changes in amount (1), C (2), and O contents (3) of epicuticular wax with branch height in *Cryptomeria japonica* (a) and in *Thujopsisu dolabrata* (b) collected in Mutsu, Aomori Prefecture (redrawn based on Sase et al. (1998a)). Plots show means and standard deviations of triplicate analyses. FL: fresh leaves. Different letters beside the plots indicate significant (p < 0.05) differences.

Leaves near the treetops may suffer more significant water stress than those on lower branches because of a decrease in water potential with increasing branch height (Zimmermann, 1971). *Cryptomeria japonica* is recognized to be more sensitive to water stress than other confers in Japan because of the high water flow resistance of the conductive tissue (Matsumoto et al., 1992). Moreover, factors in the treetop environment such as strong wind and direct sunlight may accelerate water loss, directly or indirectly. Therefore, the remarkable increase in the amount of wax with increasing branch height in *C. japonica* may be an effect of its high sensitivity to water stress, since water deficiency often stimulates wax production. In fact, the increase rate in wax amount with increasing branch height was negatively correlated with the rain factor at the sampling locations (Sase et al., 1998a). Moreover, the wax increment on higher branches may have consisted mainly of the O-rich constituents of wax (see Figure 4).

3.4. Noxious Gases

The properties of epicuticular wax for the tree growing in the volcanic area (Osorezan in Mutsu) are also shown along with those of the tree in the reference area in Figure 4. The amount of wax from the *C. japonica* tree exposed to volcanic gases was greater than that from the reference tree. The O content was significantly higher in the volcanic area than in reference area, while the C content did not show clear tendency.

The wax amount of bamboo (*Sasa kurilensis*) collected at locations close to fumaroles was also higher (16.0 $\pm$ 3.9 mg g-FL^{-1}, n = 10) than in the reference area (8.6 $\pm$ 0.8 mg g-FL^{-1}, n = 6) (Sase et al., 1998a). *Cryptomeria japonica* trees at Osorezan are continuously exposed to volcanic acidic gases (mainly H_2S). The increase in the amount of epicuticular wax in *C. japonica* leaves collected in the volcanic area may be due to exposure to volcanic acidic gases (Sase et al., 1998a). The enhanced production of epicuticular wax in *C. japonica* leaves was observed also near the electrochemical plant in Yakushima Island, and the O contents of wax increased at sites close to the electrochemical plant (Sase et al., 1998b). This phenomenon may be a physiological response to the stress caused by air pollutants including SO_2 and particulate matters from the plant. In fact, near the electrochemical plant, relatively large amounts of non-terrigenous Fe, Co, Mn, and V were found in the particulate matters on the leaves (Takamatsu et al., 2000). Therefore, it is suggested that noxious gases (such as volcanic acidic gases and anthropogenic SO_2) may stimulate the production of epicuticular wax, especially the components with a low C/O ratio, in *C. japonica* leaves.

4. DEGRADATION OF LEAF SURFACE PROPERTIES AND ITS RELATION TO TREE DECLINE

In the Kanto Plain around Tokyo, degradation of leaf surface properties may be one of significant factors causing the decline of *C. japonica*. Figure 5 shows the amount of epicuticular wax collected from three areas in Kanto District. The amount of wax in 0-y leaves showed signs of an increase with increasing levels of decline (from the mountainous area to Saitama), although the trend was not statistically significant. The reduction in the amount of wax from 0-y to 1-y leaves was larger in the Kanto Plain than in the mountainous area. This tendency was the same when the amount of wax was expressed on a leaf surface area basis (Sase et al., 1998b).

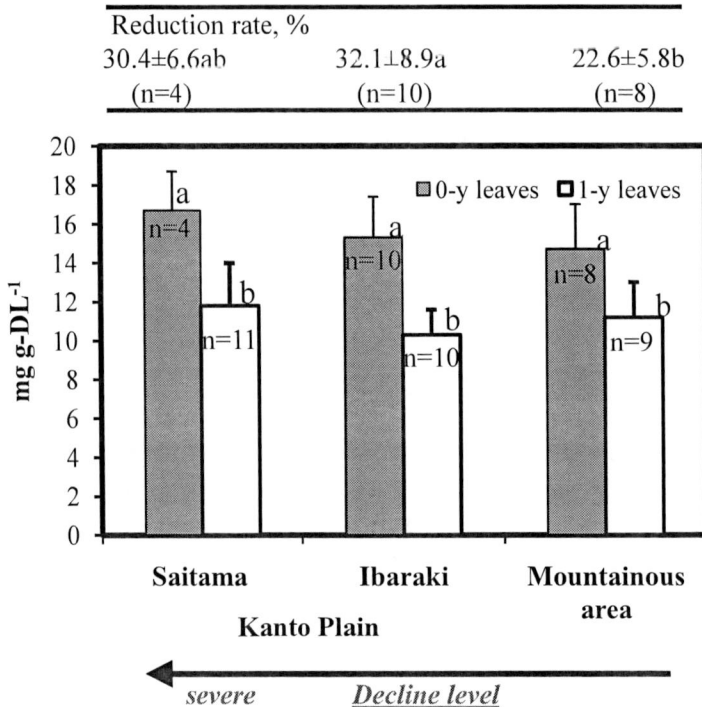

Figure 5. Reduction in amount of epicuticular wax with leaf age in *Cryptomeria japonica* from Kanto District (redrawn based on Sase et al. (1998b)). Reduction rate was calculated based on difference in wax amount between 0- and 1-year leaves. Error bar shows standard deviation. Different letters beside the values and bars indicate significant (p < 0.05) differences.

As discussed above, air pollution may stimulate the production of wax. In the Kanto Plain also, the wax production might be stimulated. Moreover, there was a positive correlation between the amount of wax in 0-y leaves and the eroded wax amount during a year, and the slope of a regression line for the Kanto Plain was steeper than that of a line for the mountainous area (Sase et al., 1998b). These results indicate that erosion of wax was faster in the Kanto Plain than in the mountainous area, even though the initial production of wax may have been stimulated in the Kanto Plain.

The degradation of wax has been observed in *Picea abies* (Grill and Golob, 1983) and in *Pinus sylvestris* (Crossley and Fowler, 1986) due to the effect of dust. In fact, amounts of anthropogenic elements such as antimony (Sb) in particulate matters deposited on the leaf surface were greater (10 times greater in case of Sb) in Saitama than in the mountainous area (Takamatsu et al., 2001a). In addition, the area of *C. japonica* decline coincides well with the distribution of high oxidant (Takahashi et al., 1987). Atmospheric deposition including the particulate matters and oxidants may be a possible cause of the wax degradation.

The degradation of wax may accelerate the cuticular transpiration rate of *C. japonica*. A negative correlation was found between the wax amount and the cuticular transpiration rate.

Moreover, the transpiration rate was relatively high in a declining area compared with the mountainous area, even though the wax was equivalent in quantity (Sase et al., 1998b). The transpiration rate in 1-y leaves was higher in the declining area (0.92% h^{-1} as decreasing rate of leaf water content) than in the mountainous area (0.60% h^{-1}).

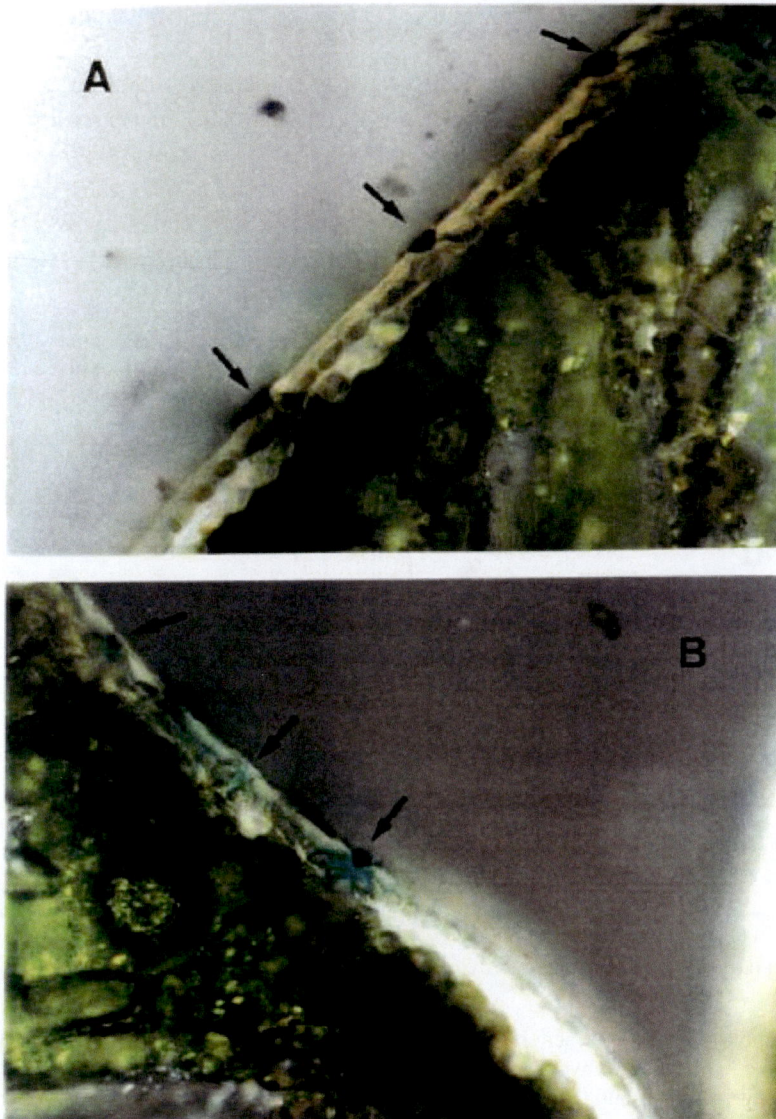

Figure 6. Penetration of dye through guard cells of stomata after abscisic acid treatment in *Cryptomeria japonica* from a declining area. Arrows indicate stomata. A: healthy stomata of a leaf collected in a mountainous area of Kanto District. B: unhealthy stomata of a leaf collected in Saitama Prefecture, a declining area.

The particulate matters could damage stomatal function, too. Since *C. japonica* has depressed stomata, particulate matters are easily trapped there and are hard to wash out. Stomata clogged with particulates were observed by scanning electron microscopes (Sase et al., 1998b; Takamatsu et al., 2001a). Figure 6 shows an example of stomatal malfunction. The leaves were dipped in an abscisic acid solution to make the stomata close and then immersed in a methylene blue solution (Takamatsu et al., 2001a). In the leaves collected in a declining area, the dye had sometimes penetrated into substomatal cavities (see Figure 6B), which suggests a depression in stomatal function (incomplete closure). In fact, the amount of particulate Sb on the leaves and the fractions of unhealthy stomata showed a good correlation (Figure 7a), suggesting that particulates on the leaves are largely responsible for the stomatal unhealthiness (Takamatsu et al., 1998b).

The stomatal malfunction may have also caused the increase in transpiration rate. Figure 7b shows the relationship between the cuticular (uncontrolled) transpiration rates and the values calculated from a linear binominal function (based on multiple regression analysis) that includes the amount of wax and particulate Sb as variables. The transpiration rates and values show a fairly good correlation. The large amount of particulate matters in the declining area may have accelerated the erosion and deterioration of the wax and caused partial malfunctioning of stomata, resulting in the increase in transpiration rate. This water loss may be a significant factor causing the decline of Japanese cedar. Moreover, in the urban area around Tokyo, the temperature is increasing, while precipitation has decreased year by year, especially since the 1950s (Matsumoto et al., 1992; Sakata, 1996; Sase et al., 1998b). The recent dry atmospheric conditions in the plain area may accelerate the water stress of trees.

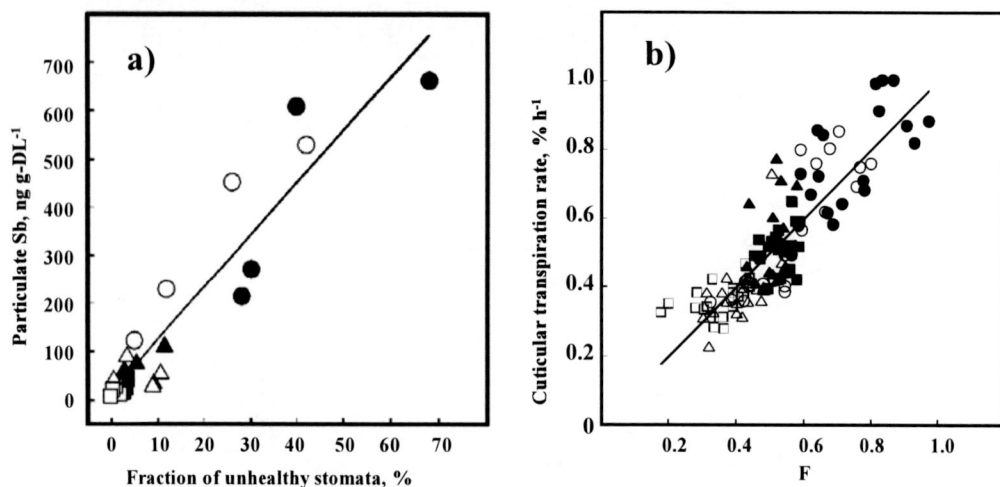

Figure 7. Relationship between fraction of unhealthy stomata and amount of Sb in particulate matters deposited on the leaves (a: redrawn based on Takamatsu et al. (2001a)) and relationship between cuticular transpiration rates and the values (F) calculated from a linear binomial function including amounts of epicuticular wax and Sb on the leaves as variables (b: redrawn based on Takamatsu et al. (2001b)). $F = 0.861 - 0.0305a + 0.000829b$, a: wax amounts (mg g-DL^{-1}), b: amounts of particulate Sb (ng g-DL^{-1}). Open and closed symbols are 0-y and 1-y leaves, respectively. Circles indicate plain areas of Saitama (and part of Tokyo), in which *C. japonica* is severely declining.

5. Particulate Matters on the Leaf Surface as a Possible Indicator of Air Pollution

Anthropogenic elements in the particulate matters on the leaf can be utilized as indicators of air pollution, too. Figure 8 shows the relationship between the amount of particulate Sb on *C. japonica* and the concentrations of NO_X or the population densities in the areas where the leaf samples were collected.

The amounts of Sb correlated with NO_X concentration and population density in each sampling area, indicating that particulate Sb originated from general human activities (Takamatsu et al., 2000). Soot deposited on the inner walls of car exhaust pipes was very rich in Sb (Takamatsu et al., 2000). Recently, Iijima et al. (2007) indicated that the properties of automotive brake abrasion dusts were consistent with the characteristics of Sb-enriched fine airborne particulate matter. Thus, transportation including fuel combustion and brake pad abrasion may be the most likely sources of the particulate Sb.

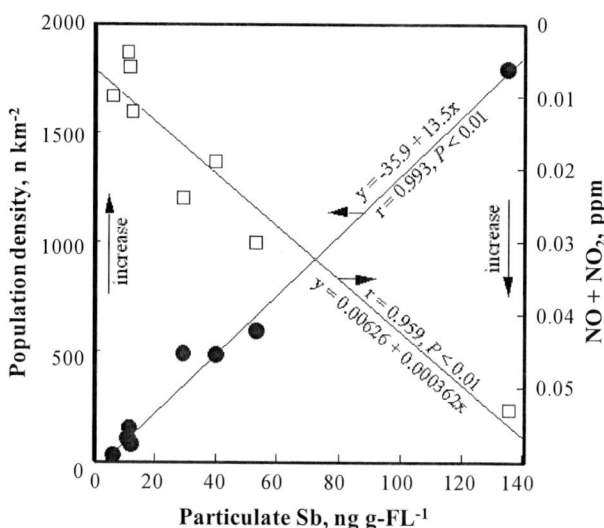

Figure 8. Relationship between average amounts of particulate Sb on 1-y leaves of *Cryptomeria japonica* and population densities (closed circles) or NO_X concentrations (open squares) in the area where the leaf samples were collected (redrawn based on Takamatsu et al. (2000)).

6. Canopy Interactions and Throughfall Chemistry

Degradation of the wax and particulate matters on the leaf surface may also affect leaf wettability, which may regulate ion exchange reactions occurring on the canopy. As a result, the changes in leaf surface properties due to air pollution are likely to induce significant changes in TF chemistry. The seasonal changes in TF and SF chemistry and the canopy interactions of K^+ and N compounds were studied in a small catchment of a *C. japonica* forest in Shibata City, Niigata Prefecture, which faces the Sea of Japan and is largely affected by seasonal winds from the west or northwest in winter (Sase et al., 2008).

The fluxes of most ions, including non-sea-salt SO_4^{2-}, from TF, SF, and RF showed distinct seasonal trends, increasing from autumn to winter, owing to the seasonal west wind, while no clear seasonal trend was observed for the fluxes of NO_3^-, NH_4^+ and K^+ ions from TF+SF, suggesting the canopy interactions (Sase et al., 2008). The mean annual flux of NH_4^+ from TF+SF was lower than that from RF, suggesting an uptake and/or consumption of NH_4^+ by the canopy. On the other hand, the substantially higher annual fluxes (> 300%) of K^+ from TF+SF than RF were attributable to leaching from the leaf surface (Sase et al., 2008). The surface properties of the leaves may affect the processes above.

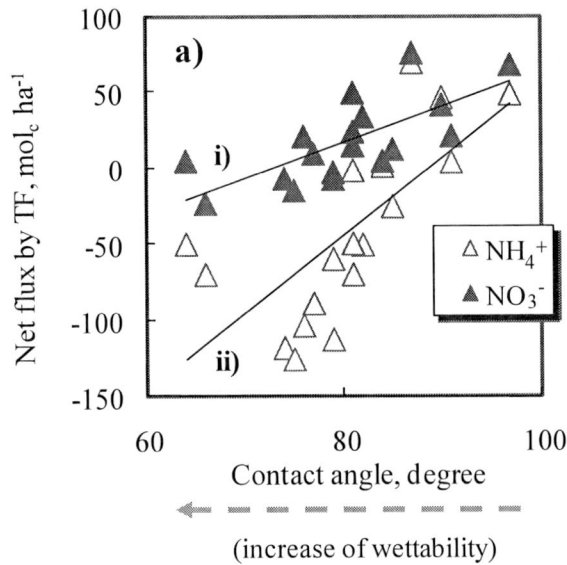

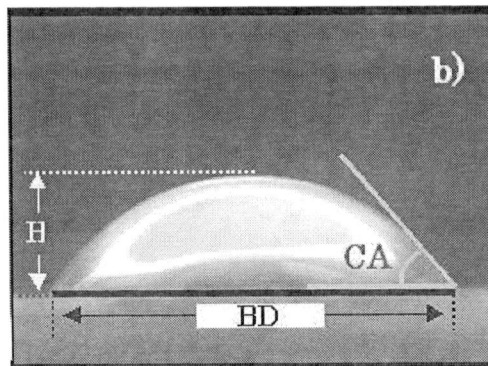

Figure 9. a) Relationship between leaf wettability and net flux of NO_3^- and NH_4^+ by throughfall and b) a typical image of the contact angle of water droplet on a *Crypromeria japoinca* leaf in November (redrawn based on Sase et al., (2008)). Regression lines represent significant correlations for (i) net flux of NO_3^- (r = 0.723, p = 0.001), and (ii) net flux of NH_4^+ (r = 0.810, p < 0.001). Conrtact angle, CA was calculated based on the height (H) and basal diameter (BD) of the droplet according to the following equation. $CA = 2\tan^{-1}[H/(BD/2)]*(180/\pi)$.

Figure 9a shows the relationship between leaf wettability of 1-y leaves and net flux of NO_3^- and NH_4^+ from TF. The net fluxes of NO_3^- and NH_4^+ decreased with increase of leaf wettability, and the correlation coefficient was larger for NH_4^+ than for NO_3^-. The concentration of K^+ increased with increase of wettability, while the concentration of NH_4^+ decreased (Sase et al., 2008). Figure 9b shows a typical image of the wettability of *C. japonica* leaf in late autumn. The wettability of leaf surface increased with leaf age (Takamatsu et al., 2001b), while the wax amounts on leaves gradually decreased, as discussed above.

The flux of K^+ from TF+SF could potentially pose a significant ion leaching effect from living foliage (Shibata and Sakuma, 1996), while inorganic N fluxes from TF+SF may have biological effects, such as uptake and consumption, on canopy surfaces (Lovett, 1992). The greater correlation coefficient of NH_4^+ compared to NO_3^- might reflect the higher rates of NH_4^+ uptake (Wilson and Tiley, 1998). Increase in leaf wettability may accelerate ion exchange on the leaf surface, resulting in increase of leaching of K^+ and uptake/consumption of N compounds on the forest canopy (Sase et al., 2008). The canopy interactions should be considered for discussion of elemental flows in forest ecosystems.

7. CONCLUSION

Rapid industrialization in the East Asian region has caused increases in emissions of various air pollutants. Acid deposition is still one of the major issues in terms of transboundary air pollution in this area, and thus the Acid Deposition Monitoring Network in East Asia (EANET) began regular-phase activities in 2001.

Forest canopy is an important interface between atmosphere and forest ecosystems. As discussed above, the degradation of epicuticular wax and stomatal unhealthiness resulting mainly from air pollution, in combination with recent dry atmospheric condition, may have placed *C. japonica* under chronic and sometimes fatal water stress in the Kanto Plain around Tokyo (Takamatsu et al., 2001a). High ozone concentration and its effects on plants is one of the hot topics in the East Asian region. Airborne particulate matters are also highlighted. Changes in leaf surface properties and their roles in East Asian forests should be considered to discuss effects of the air pollutants on ecosystems.

Moreover, nitrogen depositions and their effects on biogeochemical cycles may be another hot topic in East Asia. Annual dissolved inorganic nitrogen input in RF was approximately 18 kg N ha^{-1} y^{-1} in the small catchment in Niigata Prefecture, which exceeded thresholds in Europe and the United States (Sase et al., 2008; Kamisako et al., 2008). The magnitude of the nitrogen deposition may contribute to the high NO_3^- concentrations in the stream water and the temporary acidification that was observed during the rain events (Kamisako et al., 2008). Precise calculations of nitrogen flux and the budget in the ecosystems may need to consider canopy interactions, including leaching, uptake (or consumption), and dry deposition processes (deposition and washout). Small changes in leaf surface properties on the forest canopy may affect plant physiology and biogeochemical cycles of elements in forest ecosystems.

ACKNOWLEDGMENTS

The studies above were supported financially by the Global Environmental Research Fund (C-2, C-052, or C-082) of the Ministry of the Environment, Japan or the Grant-in-Aid for Scientific Research (No. 16510020).

REFERENCES

Crossley, A., and Fowler, D. 1986. The weathering of Scots pine epicuticular wax in polluted and clean air. New Phytol. 103: 207–218.

DeLucia, E.H., Day, T.A., and Vogelman, T.C. 1992. Ultraviolet-B and visible light penetration into needles of two species of subalpine conifers during foliar development. Plant Cell Environ. 15: 921-929.

Grill, D., and Golob, P. 1983. SEM-investigations of different dust depositions on the surface of coniferous needles, and the effect on the needle-wax. Aquilo Ser. Bot. 19: 255–261.

Günthardt-Goerg, M.S. 1994. The effect of the environment on the structure, quantity and composition of spruce needle wax. In: Air pollutants and the leaf cuticle. Springer-Verlag, Berline, Heidelberg, New York. pp. 165-174.

Iijima, A., Sato, K., Yano, K., Tago, H., Kato, M., Kimura, H., and Furuta, N. Particle size and composition distribution analysis of automotive brake abrasion dusts for the evaluation of antimony sources of airborne particulate matter. Atmos. Environ. 41 (23): 4908-4919

Kamisako, M., Sase, H., Matsui, T., Suzuki, H., Takahashi, A., Oida, T., Nakata, M., Totsuka, T., and Ueda, H. 2008. Seasonal and annual fluxes of inorganic constituents in a small catchment of a Japanese cedar forest near the Sea of Japan. Water Air Soil Pollut. 195: 51-61.

Lindberg, S.E., Lovett, G.M. 1992. Deposition and forest canopy interactions of airborne sulfur: results from the integrated forest study. Atmos. Environ. 26A, 1477-1492.

Lovett, G.M. 1992. Atmospheric deposition and canopy interactions of nitrogen, in: Johnson, D.W., Lindberg, S.E. (Eds.), Atmospheric deposition and forest nutrient cycling. Springer-Verlag, New York. pp. 152-166.

Lovett G.M., Lindberg SE, Richter DD, and Johnson DW. 1985. The effects of acidic deposition on cation leaching from three deciduous forest canopies. Can. J. For. Res. 15, 1055-1060.

Matsumoto, Y., Maruyama, Y., and Morikawa, Y. 1992. Some aspects of water relations on large *Cryptomeria japonica* D. Don trees and climatic changes on the Kanto Pains in Japan in relation to forest decline (in Japanese) Jpn. J. For. Environ. 34: 2-13.

Nashimoto, M., Takahashi, K., and Ashihara, S. 1993. Comparison of decline of Japanese cedar (*C. japonica* D. Don) stands with soil chemical properties in the Kanto-Koshin District. (In Japanese). Environ. Sci. 6: 121-130.

Percy, K.E., McQuattie, C.J., and Rebbeck, J.A. 1994. Effects of air pollutants on epicuticular wax chemical composition. In: Air Pollutants and the Leaf Cuticle. Edited by K.E. Percy, J.N. Cape, R. Jagels, and C.J. Simpson. Springer-Verlag, Berlin. pp. 67–79.

Sakata, M. 1996. Evaluation of possible causes for the decline of Japanese cedar (*Cryptomeria japonica*) based on elemental composition and δ ^{13}C of needles. Environ. Sci. Technol. **30**: 2376–2381.

Sase, H., Takamatsu, T. and Yoshida, T. 1998a. Variation in amount and elemental composition of epicuticular wax in Japanese cedar (*Cryptomeria japonica*) leaves associated with natural environmental facors. Can. J. For. Res. 28: 87-97.

Sase, H., Takamatsu, T., Yoshida, T. and Inubushi, K. 1998b. Changes in properties of epicuticular wax and the related water loss in Japanese cedar (*Cryptomeria japonica*) affected by anthropogenic environmental factors. Can. J. For. Res. 28: 546-556.

Sase, H, Takahashi, A, Sato, M, Kobayashi, H, Nakata, M, and Totsuka, T. 2008. Seasonal variation in the atmospheric deposition of inorganic constituents and canopy interactions in a Japanese cedar forest. Environ. Pollut. 152: 1-10.

Shibata, H. Sakuma, T. 1996. Canopy modification of precipitation chemistry in deciduous and coniferous forests affected by acidic deposition. Soil Sci. Plant Nutr. 42, 1-10.

Takahashi, K., Okitsu, S., and Ueda, H. 1987. Decline of Japanese cedar in Kanto-Koshin regions in relation to distributions of secondary air pollutants. (in Japanese). Trans. Annu. Meet. Jpn. For. Soc. 98: 177-180.

Takamatsu, T., Sase, H. and Takada, J. 2001a. Some physiological properties of *Cryptomeria japonica* leaves from Kanto, Japan: potential factors causing tree decline. Can. J. For. Res. 31: 663-672.

Takamatsu, T., Sase, H., Takada, J. and Matsushita, R. 2001b. Annual changes in some physiological properties of *Cryptomeria japonica* leaves from Kanto. Water Air Soil Pollut. 130: 941-946.

Takamatsu T., Takada, J., Matsushita, R. and Sase, H. 2000. Aerosol elements on tree leaves – Antimony as a possible indicator of air pollution–. Global Environ. Res. 4(1): 49-60.

Turunen M., Huttunen S. 1990. A review of the response of epicuticular wax of conifer needles to air pollution. J. Environ. Qual. 19, 35-45.

Wilson, E.J., Tiley, C. 1998. Foliar uptake of wet-deposited nitrogen by Norway spruce: An experiment using ^{15}N. Atmos. Environ. 32, 513-518.

Zimmermann, M.H. 1971. Resistance to flow in the xylem. In: Trees – structure and function. Springer-Verlag, Berlin, and New York. pp. 190-200.

In: Forest Canopies: Forest Production, Ecosystem...   ISBN 978-1-60741-457-5
Editor: J. D. Creighton and P. J. Roney   © 2009 Nova Science Publishers, Inc.

Chapter 9

EFFECTS OF FOREST CANOPY GAPS ON LITTER MICROARTHROPOD POPULATIONS IN THE SOUTHERN APPALACHIANS

Cynthia C. Kaminski, Steve Patch, and Barbara C. Reynolds

Department of Environmental Studies, University of North Carolina
at Asheville, U.S.A.

ABSTRACT

This study explored the effects of canopy gaps and gap size on leaf litter microarthropod abundance at five sites within the experimental forest of the Coweeta Hydrologic Laboratory, in the Nantahala Mountain Range of western North Carolina. In March, 2002, five canopy gaps were formed by pulling over trees to mimic natural disturbances. Two of the gaps measured 20m in diameter and three measured 40m. At each of the five gap sites we placed four experimental plots under canopy gaps and two control plots under closed canopy located just outside the gaps. Microarthropods were collected using $15cm^2$ mesh bags filled with two grams of leaf litter. Litter bags were placed in a three-by-four pattern at each plot on February 1, 2004. Litter bags and soil data were collected every other month for two years. A generalized linear mixed model was used to investigate effects of canopy gap presence and size on the populations of collembola, oribatid mites, prostigmatid mites, and mesostigmatid mites. Microarthropod counts were typically significantly greater in control plots than in gaps, with no significant effect due to gap size.

Keywords: *collembola, oribatid, leaf litter*

1. INTRODUCTION

Intact forest canopies act as buffers for the underlying forest floor ecosystem by moderating soil temperature, soil moisture evaporation, and by filtering out sunlight. Litter fall from canopies returns essential nutrients, especially carbon and nitrogen, to the soil and varies in quality according to plant species due to unique ratios of C/N availability [1-3]. Gaps in canopy cover may result in significant short term changes in soil temperature [4, 5], moisture [4, 6-8], and light availability [9, 5]. Beyond altering the composition of flora [10, 11], these abiotic factors can also alter the underlying composition of organisms ranging from large insects of the macro-fauna [12, 13] to small mites of the meso-fauna [14-20], even impacting the microbes and nematodes of the micro-fauna [6, 20, 21]. Longevity of the alteration depends on the size of the gap [10, 11] and the frequency of gap formation [23].

Microarthropods constitute a major component of biodiversity; a single square meter of forest floor can harbor hundreds of thousands of individuals belonging to a thousand different species [24]. In the food web, microarthropods act as a trophic bridge between the micro-fauna and macro-fauna by feeding on fungi, bacteria, nematodes, and each other, while acting as prey for larger organisms such as insects and spiders. The most abundant of the microarthropods are the oribatids, a suborder of mites, and collembola, also know as 'springtails' [24]. In terms of life history traits, these taxa represent opposite ends of the spectrum with oribatids displaying k-selected traits, which are best suited for stable environments, while collembola display r-selected traits that are beneficial in unstable environments. Both may be fungivores, and since fungal activity in soil systems has a strong influence on decomposition, the feeding behavior of these microarthropods can change the local fungal composition directly, leading to indirect effects on the nutrient cycle [24]. For instance, some species of collembola exhibit strong fungal selectivity [25, 26], and have been shown to disrupt the flow of carbon in mycorrhizal mycelium [27]. A study on the interrelations of fungal mycelium with co-occurring soil biota found that in addition to soil microarthropods, the microarthropods of the leaf litter also influenced soil fungi dynamics [28]. For these reasons, microarthropods are important subjects in ecological studies focusing on the recovery of habitats after disturbances, especially large scale disturbances such as burns [29], clear cuts [14, 15], tillage [16, 30], and pesticide use [31, 32].

The impact of disturbances on microarthropod populations seems to vary with the type and degree of disturbance [16, 29, 31, 33-35], initial ecosystem composition [33, 36], and physical position of study sites [29, 36]. It is therefore difficult to establish the minimum level of disturbance required to significantly affect the microarthropod population in a given ecosystem.

Research at the Coweeta Hydrologic Laboratory (USFS, Macon Co., NC) includes studying the historical and current effects of ecological disturbances and environmental gradients on biogeochemical cycling in the southern Appalachian Mountains to help predict future effects, and influence future decisions regarding land use [37]. Previous studies from Coweeta have assessed the impact of cable logging and clear cutting on both soil and leaf litter microarthropods. Densities of litter microarthropods experienced a greater impact, compared to soil microarthropods, with a significant decrease in density and change in composition. Soil fauna populations below 5cm also experienced a change in composition, but their density increased in the clear cut areas [35]. The mean annual density of litter populations remained repressed for at least eight years post clear cut [15], but after twenty-one years, the population density in the clear cut area exceeded that in the control [14].

Recently, 20m and 40m gaps were created at Coweeta to mimic a natural disturbance, such as blow-down from a hurricane [38]. This manipulation provided an opportunity to evaluate the population of litter microarthropods after a short period of recovery from a lesser degree of disturbance than implemented in past studies and to evaluate the effects of varying gap sizes.

2. METHODS

2.1. Study Site

The Coweeta Hydrologic Laboratory is a 2185-ha USDA Forest Service facility in Macon County, North Carolina, USA (35°03'N and 83°25'W). Mean annual precipitation and temperature are 245 cm and 13 °C [37]. Logging of the basin occurred around 1900 and then in the 1930's the chestnut blight caused major forest restructuring [38]. Our study plots were located in a mid-elevation hardwood forest dominated by *Quercus* spp., *Acer rubrum* L., *Acer pensylvanicum* L., and *Liriodendron tulipifera* L., located at approximately 1000 m above sea level. Soils are primarily Ultisols and Inceptisols. No forest management has been conducted since 1934. The most common agent for natural disturbance is wind from hurricanes, which generally occur in the autumn [38]. In March, 2002, five circular gaps were formed, by pulling down trees, to mimic the effects of a hurricane. Two gaps measured 20m in diameter while three measured 40m [38].

The litterbag method as outlined by Crossley [39] was followed for studying microarthropods inhabiting leaf litter using Red Oak (*Quercus rubra*) and Red Maple (*A. rubrum*). Leaves were collected from an adjacent watershed and air dried. Fiber-glass screen litterbags measuring 15 cm^2 with 1-mm mesh were filled with approximately 2.0g leaf litter and the exact weight was recorded. In the field, six plots were placed within each of the five gap sites, for a total of 30 plots. Of the six plots per site, four were located within the gapped area as experimental plots and two were placed outside the gap as controls. At each plot, twelve litterbags were staked down with surveyor's flags in a four-by-three grid pattern.

2.2. Microarthropod Collection and Extraction

Litter bags were set out in February, 2004. One litterbag from each grid was collected every 60-75 days for a total of 12 collections. The total length of the litterbag study was 754 days, with the final collection occurring in February of 2006. Collected litterbags were placed individually in plastic bags and transported in a cooler to our lab at the University of North Carolina at Asheville. Berlese funnels were used for microarthropod extraction [39]. All extracted fauna were labeled and stored in 70% ethyl alcohol. Following microarthropod extraction, the dry litter was weighed and recorded. A dissecting microscope was used to identify, count, and sort microarthropods into the following groups: collembola, oribatid mites, prostigmatid mites, and mesostigmatid mites. All groups were described as the number of animals per gram leaf litter.

2.3. Data Analysis

Numbers of microarthropods, per gram leaf litter, were analyzed using a generalized linear mixed model. Because three of the taxa had several mean counts of 0, a negative binomial distribution with a log link function was used for the error term [40]. Site was considered as a random whole plot unit with gap size as a whole plot factor and treatment as a split plot factor. Date was considered as a random factor with error terms assumed to follow a first order autoregressive pattern. To investigate the effects of regional temperature and moisture conditions, mean daily soil temperature (^{o}C) and mean daily soil moisture (m^3/m^3) were included as linear and quadratic predictor variables in the model. These data were collected from the Wayah Bald, NC, meteorological station, which is approximately 12 miles from the study site. The interactive effects of soil temperature and mean daily soil moisture with treatment and treatment*gap size were also included in the model. Statistical calculations were conducted using PROC GLIMMIX from the SAS system [41].

3. RESULTS

Data analysis began with the second collection of litter microarthropods. Over the eleven collection dates, we experienced five occasions when individual litterbags were irretrievable on their collection date, due to the presence of a copperhead (*Agkistrodon contortrix*) near the litterbag. A total of 325 litterbags were collected and from these bags 22,271 microarthropods were sorted. The treatment (gap vs. control) significantly affected the mean counts of microarthropods for each taxa (Table 1), with control plots consistently yielding greater counts (Figure 1). The interactive effect of the size of the gaps with treatment had no significant effect on the mean counts of any taxa (Table 1).

The p-values for the quadratic effects of mean daily soil temperature and mean daily soil moisture and their interactions with treatment and gap size from the Wayah station were above 0.05 for each taxon, so they were dropped from the model. The interactive effects of soil temperature and soil moisture on counts were negative for all taxa although only significant for oribatid and prostigmatid mite estimates (Table 2). Soil temperature as a numeric predictor for microarthropod response yielded positive estimates for all taxa, but the magnitude of the estimated change was only significant for oribatid and prostigmatid mites (Table 2). The use of soil moisture as a predictor for microarthropod response also yielded positive estimates for all taxa, although only the magnitude of change in oribatid mites was found to be significant (Table 2).

4. CONCLUSION

4.1. Canopy Gap

Two years after gap formation, the annual densities of the litter microarthropod community below gapped canopy sites remained depressed compared to control sites under full canopy (Figure 1). While prior disturbance studies in the Coweeta basin focused on larger scale effects, such as clear-cuts, their findings followed a similar pattern to ours with annual densities of litter microarthropods (per gram leaf litter) in cleared areas being reduced. In the previously studied clear-cuts, the microarthropod densities were reduced by 50% or more compared to control densities [35] and 28% less eight years later [17]. These effects were partially attributed to microclimatic changes of increased temperature extremes and intensified wet/dry cycles due to the removal of cover [17]. We suspect these same microclimatic factors contributed to the differences in microarthropod densities between our treatments.

We did not find any significant differences in annual densities of litter microarthropods between the 20m and 40m gap sites, indicating that at the ecosystem level of the meso-fauna, a gap 40m in diameter is indistinguishable from one that is 20m in diameter. A one year gap study in a subtropical forest found that bacterial and fungal respirations in leaf litter under medium sized gaps (measuring between 15m and 30m in diameter) did not differ significantly from large gaps (those over 30m in diameter) [6]. However, respirations of both size classes differed significantly from those under closed canopy and remained so throughout the study [6]. If microarthropod abundances are regulated primarily by competition for food, as suggested by Ferguson et al. [20], then the next question is: what regulates food production? For bacteria and fungi, the primary food source of oribatid mites and collembola [25], the major limiting factors for growth are temperature and moisture [6, 20, 21].

4.2. Soil Moisture and Temperature

Our model of predicted estimates showed a positive relationship between Wayah soil moisture and abundance of all taxa, but was only significant for oribatid mites (Table 2). While both oribatid and collembola populations are immediately reduced by drought conditions [9], collembola tend to recover faster than oribatids after disturbance due to their higher turnover rates. An induced drought study in Norway resulted in eight species of collembola recovering within one year while all negatively affected species of oribatids remained significantly reduced [9]. In our study, the gap sites had a two-year recovery before data collection began, resulting in substantial vegetative succession, both from sprout regrowth and young American chestnut trees (*Castanea dentata*). This regrowth may have reduced the gap effect in terms of both soil moisture and temperature. Additionally, the longevity of the initial gap effect may have been reduced in some gap plots due to leaf litter inputs from surrounding intact canopy. The relatively small size of gaps and the sloping topography of our sites make this especially likely in the 20m gaps.

Emigration of microarthropods away from drought conditions has been demonstrated locally in both the soil layer and the litter layer by Seastedt and Crossley [35]. They found that the proportion of total soil microarthropod populations in the upper section of soil cores from a clear cut watershed was lower than in the control cores [35], indicating either a vertical or horizontal migration away from adverse conditions. On a broader scale, collembola and oribatid densities have displayed significant, positive relationships with soil moisture in a wide array of habitats, including Kentucky oak-maple forests [8], tropical dry forests in western Mexico [19], and rain forests in Nigeria [18].

In terms of Wayah soil temperature, our model also showed a positive relationship for all taxa, with significance for both oribatid and prostigmatid mites (Table 2). This relationship is supported by a study conducted in Saskatchewan, Canada [20], but counter to the findings by Seastedt and Crossley, who reported soil temperatures exceeding lethal values to oribatid mites (>40 C) in their clear-cut areas [35]. However, our sites experienced a smaller scale disturbance than the previous Coweeta study and did not result in soil temperatures exceeding 27 C. The results of the litter study in western Mexico also countered ours, but again the average temperature in the western Mexico site is double that in the Coweeta basin [19] and soil temperatures in Coweeta can drop down to zero C in winter months. If temperature ranges in Coweeta represent the lower end of the tolerance range for most microarthropods, and temperature ranges in Mexico represent the upper tolerance range, then it would make sense that the temperature relationships are reversed.

When the interactive effects of Wayah soil temperature and Wayah soil moisture were considered, the relationship for all taxa became negative (Table 2), which means combinations of either low temperature-low moisture or high temperature-high moisture result in lower microarthropod counts than those expected if the combination is one of high-low. This becomes important if we consider both the direct and indirect effects of habitat alteration on microarthropods. Such alterations have been shown to affect microarthropods directly through inducing vertical movement into the soil [17, 35], but also indirectly by altering local fungal and bacterial growth [6].

4.3. Future Research

Microarthropods operate in microclimates that can vary greatly within a single region and absolutely vary between regions. However, much of the information on microarthropods and soil ecology originates from multiple studies evaluating one or two factors of an ecosystem over multiple climatic regions. Few comprehensive ecological studies have been conducted to fully evaluate the soil interactions within a single region, which forces scientists to compare their results to a diverse set of regions that may have little in common with theirs. We were fortunate that previous studies were conducted in the same region and for similar purposes as ours. However, our study would have been improved by obtaining more data on the local temperature and moisture conditions for direct comparisons between treatments. Also, we analyzed populations of microarthropods by order and suborder, but different species of both collembola and oribatids have shown various tolerances to drought according to reproductive mode [7] as well as feeding preferences that have long term implications for the local fungal community composition [25, 28]. Likewise, initial fungal composition may drive microarthropod community composition [24], so identifying and monitoring changes in fungal community due to disturbance would offer a more comprehensive insight. When possible, scientists of different disciplines should share study sites to create comprehensive regional profiles and then compare relationships to discover global trends.

5. TABLES AND FIGURES

Table 1. Microarthropod responses to plot treatments. Treatment was either control (n =99) or gap (n=226), while Gap Size x Treatment was either 20m (n_C=88, n_G=41) or 40m (n_C=55, n_G=141) in diameter

Response	Treatment (p-value)	Gap Size x Treatment (p-value)
Collembola	0.006	0.190
Oribatid	0.007	0.274
Prostigmatid	0.029	0.709
Mesostigmatid	0.041	0.339

Table 2. Microarthropod responses to soil characteristics. Because a log link function was used on the data, +/- indicates the direction of the predicted change in mean count

Response	Soil Temperature Est. +/- (p-value)	Soil Moisture Est. +/- (p-value)	Interactive Effect Est. +/- (p-value)
Collembola	+ (0.099)	+ (0.112)	- (0.056)
Oribatid	+ (0.002)	+ (0.003)	- (0.002)
Prostigmatid	+ (0.039)	+ (0.069)	-(0.021)
Mesostigmatid	+ (0.271)	+ (0.614)	- (0.524)

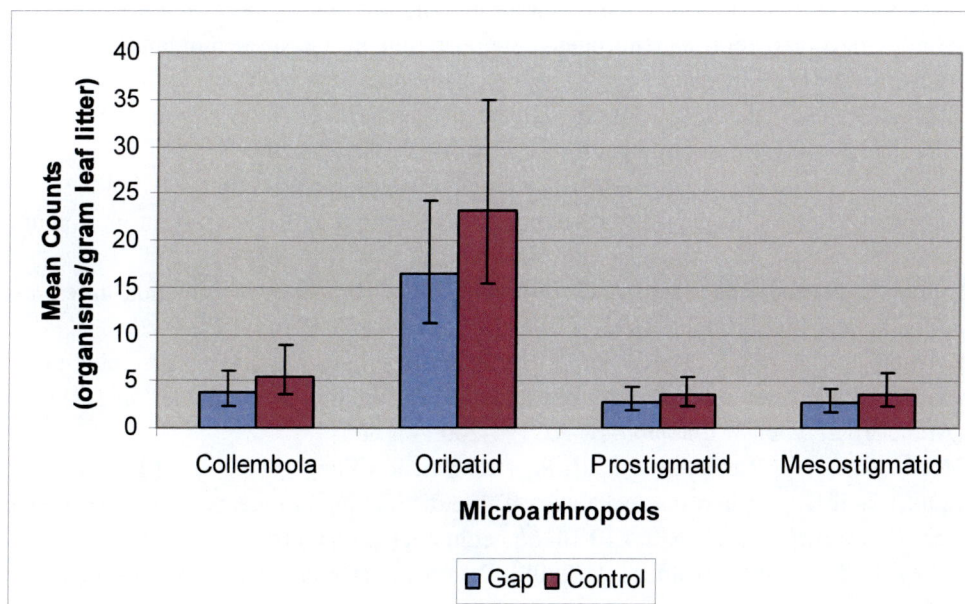

Figure 1. Mean Microarthropod Counts.

The counts of the four categories of microarthropods were pooled across collection dates, according to plot treatment (gap or control). The estimated means at mean soil temperature and soil moisture conditions were then reversed transformed from the negative binomial model and graphed as mean number of organisms per gram dry leaf litter. The bars are confidence intervals. For all categories, Gap n=226 and Control n=99.

ACKNOWLEDGEMENTS

We would like to thank James S. Clark, Duke University, for initiating the gap project, and Phaedra Scarborough for her initial work on the project. We would also like to thank the National Science Foundation, Grant DEB – 0218001, for funding this study.

REFERENCES

[1]   Zhang, P., Tian, X., He, X., Song, F., Ren, L., and Jiang, P. (2008). Effect of litter quality on its decomposition in broadleaf and coniferous forest. *European Journal of Soil Biology,* doi:10.1016/j.ejsobi.2008.04.005.
[2]   Knoepp, J. D., Reynolds, B. C., Crossley, D. A., and Swank, W. T. (2005). Long-term challenges in forest floor processes in southern Appalachian forests. *Forest Ecology and Management, 220,* 300-312.
[3]   Sariyildiz, T. and Anderson, J. M. (2005). Variation in the chemical composition of green leaves and leaf litters from three deciduous tree species growing on different soil types. *Forest Ecology and Management, 210,* 303-319.
[4]   Prescott, C. E. (1997). Effects of clearcutting and alternative silvicultural systems on rates of decomposition and nitrogen mineralization in a coastal montane coniferous forest. *Forest Ecology and Management, 95,* 253-260.
[5]   Zhang, Q. and Zak, J. C. (1995). Effects of gap size on litter decomposition and microbial activity in a subtropical forest. *Ecology, 76,* 2196-2204.
[6]   Zhang, Q. and Zak, J. C. (1998). Potential Physiological Activities of Fungi and Bacteria in relation to plant litter decomposition along a gap size gradient in a natural subtropical forest. *Microbial Ecology, 35,* 172-179.
[7]   Lindberg, N. and Bengtsson, J. (2005). Population responses of oribatids mites and collembolans after drought. *Applied Soil Ecology, 28,* 163-174.
[8]   Lensing, J. R., Todd, S., and Wise, D. H. (2005). The impact of altered precipitation on spatial stratification and activity-densities of springtails (Collembola) and spiders (Araneae). *Ecological Entomology, 30,* 194-200.
[9]   Marthews, T. R., Burslem, D. F. R. P., Paton, S. R., Yanguez, F., and Mullins, C. E. (2008). Soil Drying in a tropical forest: Three distinct environments controlled by gap size. *Ecological Modelling,* doi:10.1016/j.ecolmodel.2008.05.011.
[10] Fahey, R. T. and Puettmann, K. J. (2008). Patterns in spatial extent of gap influence on understory plant communities. *Forest Ecology and Management, 255,* 2801-2810.
[11] Thompson, J., Proctor, J., Scott, D. A., Fraser, P. J., Marrs, R. H., Miller, R. P., and Viana, V. (1998). Rain Forest on Maraca Island, Roraima, Brazil: artificial gaps and plant response to them. *Forest Ecology and Management, 102,* 305:321.

[12] Negrete-Yankelevich, S., Fragoso, C., Newton, A. C., Russell, G., and Heal, O. W. (2008). Species-specific characteristics of trees can determine the litter macroinvertebrate community and decomposition process below their canopies. *Plant Soil, 30,* 83-97.

[13] Greenberg, C. H. and Forrest, T. G. (2003). Seasonal abundance of ground-occuring macroarthropods in forest and canopy gaps in the southern Appalachians. *Southeastern Naturalist, 2,*591-608.

[14] Heneghan, L., Salmore, A., and Crossley Jr., D. A. (2004). Recovery of decomposition and soil microarthropod communities in an Appalachian watershed two decades after a clearcut. *Forest Ecology and Management, 189,* 353-362.

[15] Blair, J. M. and Crossley, D. A. (1988). Litter decomposition, nitrogen dynamics and litter microarthropods in a southern Appalachian hardwood forest 8 years following clearcutting. *Journal of Applied Ecology, 25,* 683-698.

[16] Reeleder, R. D., Miller, J. J., Ball Coelho, B. R., and Roy, R. C. (2006). Impacts of tillage cover crop, and nitrogen on populations of earthworms, microarthropods, and soil fungi in a cultivated fragile soil. *Applied Soil Ecology, 33,* 243-257.

[17] Badejo, M. A., Nathaniel, T. I., and Tian, G. (1998). Abundance of springtails (Collembola) under four agroforestry tree species with contrasting litter quality. *Biol. Fertil Soils, 27,* 15-20.

[18] Badejo, M. A. and Akinwole, P. O. (2006). Microenvironmental preferences of oribatids mite species on the floor of a tropical rainforest. *Experimental Applied Acarol, 40,* 145-156.

[19] Palacios-Vargas, J. G., Castano-Meneses, G., Gomez-Anaya, J. A., Martinez-Yrizar, A., Mejia-Recamier, B. E., and Martinez-Sanchez, J. (2007). Litter and soil arthropods diversity and density in a tropical dry forest ecosystem in Western Mexico. *Biodivers Conserv, 16,* 3703-3717.

[20] Ferguson, S. H. and Joly, D. O. (2002). Dynamics of springtail and mite populations: the role of density dependence, predation, and weather. *Eco. Ento, 27,* 565-573.

[21] Kranabetter, J. M. and Wylie, T. (1998). Ectomycorrhizal community structure across forest openings on naturally regenerated western hemlock seedlings. *Canadian Journal of Botany, 76,* 189-196.

[22] Ritter, E. and Bjornlund, L. (2005). Nitrogen availability and nematode populations in soil and litter after gap formation in a semi-natural beech-dominated forest. *Applied Soil Ecology, 28,* 175-189.

[23] Brown, N. D. and Whitemore, T. C. (1992). Do dipterocarp seedlings really partition tropical rain forest gaps? In: Marshall, A. G., Swaine, M. D. (Eds.), *Tropical Rain Forest: Disturbance and Recovery.* (pp. 369-378). The Royal Society, London.

[24] Coleman, D., Crossley Jr., D. A., and Hendrix, P. (2004). *Fundamentals of Soil Ecology* (2nd ed.). New York: Elsevier Academic Press, 386 pp.

[25] Tordoff, G., Boddy, L., and Jones, T. H. (2008). Species-specific impacts of collembolan grazing on fungal foraging ecology. *Soil Biology and Biochemistry, 40,* 434-442.

[26] Larsen, J., Johansen, A., Larsen, S. E., Heckmann, L. H., Jakobsen, I., and Krogh, P. H. (2008). Population performance of collembolans feeding on soil fungi from different ecological niches. *Soil Biology and Biochemistry, 40,* 360-369.

[27] Johnson, D., Krsek, M., Wellington, E. M. H., Scott, A. W., Cole, L., Bardgett, R. D., Read, D. J., and Leake, J. R. (2005). Soil invertebrates disrupt carbon flow through fungal networks. *Science, 309,* 1047.

[28] Krivtsov, V., Griffiths, B. S., Salmond, R., Liddell, K., Garside, A., Bezginova, T., Thompson, J. A., Staines, H. J., Watling, R., and Palfreyman, J. W. (2004). Some aspects of interrelations between fungi and other biota in forest soil. *Mycological Research, 108,* 933-946.

[29] Dress, W. and Boerner, R. E. J. (2003). Patterns of microarthropod abundance in oak-hickory forest ecosystems in relation to prescribed fire and landscape position. *Pedobiologia, 48,* 1-8.

[30] Adl, S. M., Coleman, D. C., and Read, F. (2006). Slow recovery of soil biodiversity in sandy loam soils of Georgia after 25 years of no-tillage management. *Agriculture, Ecosystems and Environment, 114,* 323-334.

[31] Cortet, J., Gillon, D., Joffre, R., Ourcival, J. M., and Poinsot-Balaguer, N. (2002). Effects of pesticides on organic matter recycling and microarthropods in maize field: use and discussion of the litterbag methodology. *European Journal of Soil Biology, 38,* 261-265.

[32] Doles, J., Zimmerman, R. J., and Moore, J. C. (2001). Soil microarthropod community structure and dynamics in organic and conventionally managed apple orchards in Western Colorado, USA. *Applied Soil Ecology, 18,* 83-96.

[33] Cole, L., Buckland, S. M., and Bardgett, R. D. (2008). Influence of disturbance and nitrogen addition on plant and soil animal diversity in grassland. *Soil Biology and Biochemistry, 40,* 505-514.

[34] Tsiafouli, M. A., Kallimanis, A.S., Katana, E., Stamou, G. P., and Sgardelis, S. P. (2005). Responses of soil microarthropods to experimental short-term manipulations of soil moisture. *Applied Soil Ecology, 29,* 17-26.

[35] Seastedt, T. R. and Crossley Jr., D. A. (1981). Microarthropod response following cable logging and clear-cutting in the southern Appalachians. *Ecology, 62,* 126-135.

[36] Jenson, P., Jacobson, G. L., and Willard, D. E. (1973). Effects of mowing and raking on collembolan. *Ecology, 54,* 564-572.

[37] National Science Foundation (NSF) and University of Georgia. (2008) Coweeta Long Term Ecological Research. Available from: <http://cwt33.ecology.uga.edu/research/contents_research.htm>.

[38] Dietze, M. C. and Clark, J. S. (2008). Changing the gap dynamics paradigm: vegetative regeneration control on forest response to disturbance. *Ecological Monographs, 78,* 331-347.

[39] Crossley Jr., D. A. and Blair, J. M. (1991). A high efficiency, "low-technology" Tullgren-type extractor for soil Microarthropods. *Agriculture, Ecosystem and Environment, 34,* 187-192.

[40] Sileshi, G. (2008). The excess-zero problem in soil animal count data and choice of appropriate models for statistical inference. *Pedo biologia, 52,* 1-17.

[41] SAS Institute. SAS/STAT Users Guide.http://support.sas.com/onlinedoc/913/docMainpage.jsp

In: Forest Canopies: Forest Production, Ecosystem…   ISBN 978-1-60741-457-5
Editor: J. D. Creighton and P. J. Roney   © 2009 Nova Science Publishers, Inc.

Chapter 10

INTERACTIONS BETWEEN URBAN VEGETATED SURFACES AND THE ATMOSPHERE

Timo Vesala[1], Leena Järvi[1], Üllar Rannik[1], Sampo Smolander[1], Andrey Sogachev[2], and Eero Nikinmaa[3]

[1] Department of Physics, University of Helsinki, Finland
[2] Wind Energy Division, Risø National Laboratory for Sustainable Energy, Technical University of Denmark – DTU, Roskilde, Denmark
[3] Department of Forest Ecology, University of Helsinki, Finland

ABSTRACT

Within the framework of micrometeorology and biosphere-atmosphere interactions, one of the important scientific tasks has recently been to conduct long-term flux measurement sites in an array of land biomes and climates world-wide in order to gain understanding of exchange processes with the good spatio-temporal coverage. At the moment, the flux network is rather dense for natural and seminatural ecosystems while only a few measurements have been conducted over urban and sub-urban landscapes. Urban vegetated surfaces do not likely contribute much to overall material balances, like carbon sinks/sources, but locally they may be important; they have a significant role in the well-being of the population, and knowledge of them is scarce. Besides carbon, vegetation also affects the energy balance by modifying surface temperature. Canopies may also act as significant sinks for aerosol particles thus cleaning the air, although vegetation can also produce particles. Apart from telling the impacts that urban vegetation have on the local air properties, the flux studies provide valuable information on the performance of the vegetation, which is valuable knowledge for the maintenance of green surfaces in urban areas. We discuss the present status of urban flux studies from the point of view of vegetated surfaces and try to point out areas that need to be explored more. A few illustrative results on flux studies from Helsinki, Finland, are presented.

INTRODUCTION

The biosphere interacts with the atmosphere and the interaction is revealed in the form of the exchange of mass, heat and momentum between those systems. Quantitatively the amount of exchange is expressed as the exchange rate, which is the flux (density) of the material or energy giving the amount of the material or energy exchanged by the land surface and the atmosphere per unit area and unit time. This interaction is mediated by atmospheric turbulence within the atmospheric boundary layer and especially within the lowest part, the surface layer, being typically 100–200 m over the land (Foken 2008). The interaction is acting in both directions: for example, the surface vegetation substantially affects the microclimate within the plant community and the surface layer, and, in turn, the microclimate influences the plant community. The primary method adopted for CO_2 and H_2O flux measurements is presently a micrometeorological direct technique called eddy covariance (EC), measuring flux densities directly between biosphere and atmosphere typically from towers (Baldocchi et al., 2001). However, EC methodology still possesses some critical issues under debate and it is important that exchange processes are also monitored by means of other methods, such as chamber measurements. EC measurements are quite complex and need comprehensive micrometeorological knowledge (see e.g. Foken 2008; Aubinet et al. 2000; Finnigan 2008).

Within the framework of micrometeorology and biosphere-atmosphere interactions, one of the important scientific tasks has recently been to establish long-term EC flux measurement sites in an array of land biomes and climates world-wide. At the moment, the network is rather dense for natural and seminatural ecosystems while only a few measurements have been conducted over urban and sub-urban landscapes. Urban vegetated surfaces do not likely contribute much overall material balances, like carbon sinks/sources, but locally they may be important, and knowledge of them is scarce. Besides carbon, vegetation also affects the energy balance modifying surface temperature and changing the fraction of the available energy consumed to heating of the surface vs. consumed to the evapo-transpiration. Canopies may also act as significant sinks for aerosol particles thus cleaning the air, although vegetation can also produce particles (Kulmala, 2003). Only a few studies have been carried out on emission/deposition fluxes of aerosol particles over urban surfaces (Dorsey et al., 2002; Mårtensson et al., 2006; Schmidt and Klemm, 2008).

ON EDDY COVARIANCE FLUX MEASUREMENT TECHNIQUE

The EC technique facilitates direct turbulent flux measurements without affecting the natural gas transfer between the surface and air, since the EC set-up (basically a gas analyzer and an anemometer) is installed above the surface in a tower. It also provides a tool to estimate exchange over larger areas than, for example, a single measuring chamber. However, whereas a chamber confines a known source area for its measurement, the area represented by EC measurements, the footprint, is a complex function of the observation level, surface roughness length and canopy structure together with meteorological conditions (wind speed and direction, intensity of turbulence and atmospheric stratification). The footprint defines the field of view, reflecting the influence of the surface on the measured turbulent flux (or concentration). Most often, fluxes are calculated using ½–1 hr time averages and each single flux value represents a different source area, although the difference between the areas corresponding to two successive values may not be large.

Urban environments are typically very heterogeneous and the roughness elements (such as buildings) can be high. This may often lead to the situation where classical micrometeorological preconditions of flux measurements (Roth, 2000) are compromised with practical limitations, especially, if the measurements are carried out within the roughness sub-layer close to the canopy. However, the limiting application conditions of the methods in areas with obstacles and in complex terrain have been recently relaxed with the more standardized quality assessment routines of the data (Foken 2008). All in all, the amount of micrometeorological studies and their applications are expanding and getting new scientific features. Many of the problems and challenges are interrelated and they are at the center of inquiry in the "new" micrometeorology of the real world (see Schmid, 2002), like the complex urban areas.

EDDY COVARIANCE FLUX STUDIES OVER URBAN SURFACES

The longest CO_2 flux record obtained by EC over the urban surface is reported by Soegaard and Møller-Jensen (2003), who had measured fluxes in the center of Copenhagen since 2000. Earliest measurements go back in the summer of 1995 (Grimmond et al., 2002), but the campaign undertaken in Chicago was only about 2 months. Other studies concern Southern (Grimmond et al., 2004; Salmond et al., 2005), Central (Nemitz et al., 2002; Vogt et al., 2006; Schmidt et al., 2008) and Northern European sites (Vesala et al., 2008), Japanese locations (Moriwaki et al., 2004; Moriwaki and Kanda, 2006), Mexico City (Velasco et al., 2005) and Melbourne (Coutts et al., 2007). In one of the most recent papers, Vogt et al. (2006) came to the conclusion that the diversity of urban areas is not yet adequately covered by experimental studies and more long-term studies from a variety of cities are needed.

FLUX MEASUREMENT RESULTS FROM KUMPULA, HELSINKI

In Kumpula, Helsinki, at the SMEAR (Station for Measuring Ecosystem–Atmosphere Relationships) III station (60°20' N, 24°96' E), the flux tower represents a typical urban surface. In one wind direction the fetch is covered by vegetation while in the other direction passes one of the main roads to Helsinki centre with heavy traffic loads. The third main wind sector contains mostly buildings and small roads including a parking lot. The sensible heat flux (heat flux associated to the vertical movement of warmer/cooler air) (H) and momentum flux (vertical transport rate of momentum in the air, describing also the amount of the mechanical turbulence) measurements started in 2004 and the latent heat flux (vertical flux of heat from the surface by evapotranspiration) (LE) and CO_2 flux (F_c) measurements in 2005. The aerosol particle number flux has been measured since 2007. All measurements are still going on and they are planned to be very long-term. The auxiliary measurements include PAR (Photosynthetically Active Radiation) and net radiation, wind speed, air temperature and

Timo Vesala, Leena Järvi, Üllar Rannik et al.

various trace gas concentrations and aerosol particle properties. Figure 1 shows the median CO_2 flux for each day between 10 am–3 pm at the Helsinki site for almost three years.

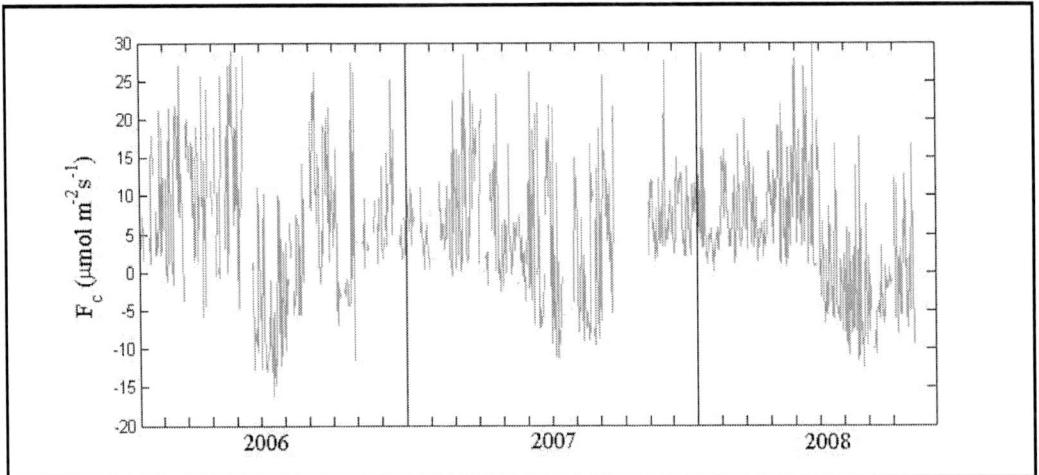

Figure 1. The flux carbon dioxide flux at each day between 10 am–3 pm at Kumpula, Helsinki, measured by the eddy covariance method. The positive values indicate upward flux (emission) and negative downward (uptake).

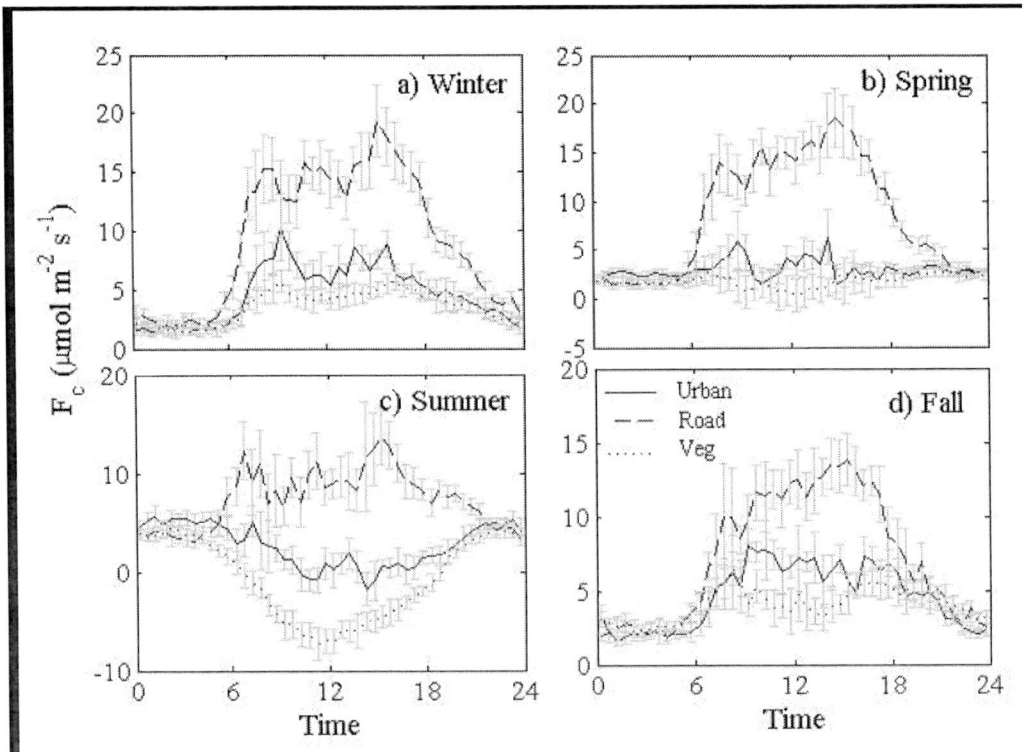

Figure 2. The median seasonal diurnal carbon dioxide flux for Dec-Feb, Mar-May, Jun-Aug and Sep-Nov over the period Jan 2006 – Aug 2008, at Kumpula, Helsinki, measured by the eddy covariance method. The positive values indicate upward flux (emission) and negative downward (uptake).

The average day-time flux is dominantly upwards (emission), but in the summer occasionally downward fluxes can also be observed (assimilation by vegetation).

For flux climatology, the fluxes have been analysed as average diurnal courses over winter, spring, summer and fall. H typically exceeds LE reaching 300 W m^{-2} over urban and road surfaces in the summer and 100 W m^{-2} in the winter. LE is highest in the summer over vegetation cover attaining 150 W m^{-2}. The emission rate of CO_2 was high over road sector up to 20 µmol m^{-2} s^{-1} while in the vegetation sector it remained below 8 µmol m^{-2} s^{-1} in the summer being occasionally even negative due to the carbon uptake by vegetation (Figure 2). The maximum single ½ hourly value of F_c was 70 µmol m^{-2} s^{-1} and the minimum – 10 µmol m^{-2} s^{-1}. F_c correlated with traffic density and a background non-vehicle flux deduced from the interception of flux-vs.-vehicle count regression was 2 µmol m^{-2} s^{-1}.

The results can be compared with earlier studies. Moriwaki et al. (2004) have collected emission data based on EC measurements from densely built-up Basel (Vogt et al., 2003), suburban Chicago (Grimmond et al., 2002) and City center Edinburgh (Nemitz et al., 2002) beside their own measurements over residential Tokyo. Highest emission rates up to 75 µmol m^{-2} s^{-1} was found in Edinburgh while the lowest maximum values of around 10 µmol m^{-2} s^{-1} were detected in Tokyo in summer and in Chicago. The lowest flux in Tokyo was about 5 µmol m^{-2} s^{-1} and in other sites about 0 µmol m^{-2} s^{-1}. Grimmond et al. (2004) reported on measurements above the center of Marseille in summer. The city district was found to be almost always a source (up to around 40 µmol m^{-2} s^{-1}) but the vegetation reduced the fluxes in the afternoon. In Copenhagen, the emission rates range from less than 5 µmol m^{-2} s^{-1} in the residential areas up to 100 µmol m^{-2} s^{-1} along the major roads in the city center (Soegaard and Møller-Jensen, 2003). The city centre of Münster in Germany appeared to be a significant source of carbon as well (Schmidt et al., 2008). Velasco et al. (2005) concluded about the results from Mexico City that the mean daily fluxes were similar to those observed in European and US cities. Finally, Coutts et al. (2007) carried out flux measurements at two sites in Melbourne, another being a suburban site and the other with differing surface characteristics, especially having greater vegetation cover. They found that summer time fluxes were lower than during winter due to greater vegetative influence and reduced natural gas combustion, but although vegetation limited the source of carbon in the afternoon, it was not enough to cancel anthropogenic emissions. The results from Helsinki regarding general level of emissions and variations resemble those reported earlier. The effect of vegetation for lowering the fluxes is distinguishable in some studies. However, at the Helsinki site the effect is most pronounced since the main vegetation-sector as a source area includes very few anthropogenic sources and thus the surface can act even as CO_2 sink in summer days. Figure 2 shows the fluxes at the Helsinki site for four seasons and for the three different land-uses.

The soil chamber measurements at the Helsinki site have indicated that the respiration level of the treeless vegetative surface was of the order of 1 - 3 µmol m^{-2} s^{-1}. The chamber measurements were close to nocturnal EC estimates.

CONCLUSIONS

Besides forests, wetlands, grasslands and arable areas, long-term flux measurements are also needed from urban sites to reveal the dynamics of greenhouse gas, energy and particle fluxes and how these might be related to variables influencing urban vegetation. The flux may depend drastically on surface type (Vesala et al., 2008). Besides CO_2, H_2O and aerosol particles the high-response (measuring concentration variations with the time resolution of 1 Hz or better) instruments suitable for eddy covariance are available to O_3, nitrogen oxides (NO_x), nitrous oxide (N_2O), CH_4, NH_3 and some hydrocarbons, but the flux studies of these compounds in the urban environment are still very nascent.

REFERENCES

Aubinet, M., Grelle, A., Ibrom, A. Rannik, Ü., Moncrieff, J., Foken, T., Kowalski, A., Martin, P., Berbigier, P., Bernhofer, C., Clement, R., Elbers, J., Granier, A., Grünwald, T., Morgenstern, K., Pilegaard, K., Rebmann, C., Snijders, W., Valentini, R., Vesala, T. *Adv. Ecol. Res.* 2000, *vol* 30, 113–175.

Baldocchi, D., Falge, E., Lianhong, G., Olson, R., Hollinger, D., Running, S., Anthoni, P., Bernhofer, Ch., Davis, K., Evans, R., Fuentes, J., Goldstein, A., Katul, G., Law, B., Lee, X., Malhi, Y., Meyers, T., Munger, W., Oechel, W., Paw U, K.T., Pilegaard, K., Schmid, H.P., Valentini, R., Verma, S., Vesala, T., Wilson, K., Wofsy, S. *Bull. Am. Meteor. Soc.* 2001, *vol* 82, 2415-2434.

Coutts, A.M., Beringer, J., Tapper, N.J. *Atmospheric Environ.* 2007, *vol* 41, 51-62.

Dorsey, J.R., Nemitz, E., Gallagher, M.W., Fowler, D., Williams, P.I., Bower, K.N., Beswick, K.M. *Atmos. Environ.* 2002, *vol* 36, 791-800.

Finnigan, J. *Ecological Applications.* 2008, *vol* 18, 1340-1350.

Foken, T. *Micrometeorology*; Springer: Berlin Heidelberg, Germany, 2008.

Grimmond C. S. B., King, T. S., Cropley, F. D., Nowak, D. J., Souch C. *Environ. Pollution* 2002 ,*vol* 116, 243-254.

Grimmond, C. S. B., Salmond, J. A., Oke, T. R., Offerle, B., Lemonsu, A. *J. Geophys. Res.* 2004, *vol* 109, D24101, doi: 10.1029/2004JD004936.

Kulmala M. *Science* 2003, *vol* 302, 1000-1001.

Mårtensson, E.M., Nilsson, E.D., Byzorius, G., Johansson, C. *Atmos. Chem. Phys.* 2006, *vol* 6, 769-785.

Moriwaki, R., Kanda, M. *J. Appl. Meteorol.* 2004, *vol* 43, 1700-1710.

Moriwaki, R., Kanda, M. *Boundary-Layer Meteorol.* 2006, *vol* 120, 163-179.

Nemitz, E., Hargreaves, K. J., McDonald, A. G., Dorsey, J. R., Fowler, D. *Environ. Sci. Technol.* 2002, *vol* 36, 3139-3146.

Roth, M. *Quart. J. Roy. Meteor. Soc* 2000, *vol* 126, 941-990.

Salmond, J. A., Oke, T. R., Grimmond, C. S. B., Roberts, S., Offerle, B. *J. Appl. Meteorol.* 2005, *vol* 44, 1180-1194.

Schmid, H.P. *Agric. Forest Meteorol* 2002, *vol* 113, 159-183.

Schmidt, A., Klemm, O. *Atmos. Chem. Phys. Discuss.* 2008, *vol* 8, 8997-9034.

Scmidt, A., Wrzesinksy, T., Klemm, O. *Boundary-Layer Meteorol.* 2008, *vol* 126, 389-413.

Soegaard, H., Møller-Jensen, L. *Remote Sensing Environ.* 2003, *vol* 87, 283-294.

Velasco, E., Pressley, S., Allwine, E., Westberg, H., Lamb, B. *Atmospheric Environ.* 2005, *vol* 39, 7433-7446.

Vesala, T., Järvi, L., Launiainen, S., Sogachev, A., Rannik, Ü., Mammarella, I., Siivola, E., Keronen, P., Rinne, J., Riikonen, A., Nikinmaa, E. *Tellus* 2008, *vol* 60B, 188-199.

Vogt, R., Christensen, A., Rotach, M. W., Roth, M., Satyanarayana, A. N. V. In *Preprints, 5th International Conference on Urban Climate*, Lodz, Poland, International Association for Urban Climate, 2003, 321-324.

Vogt, R., Christensen, A., Rotach, M. W., Roth, M., Satyanarayana, A. N. V. *Theor. Appl. Climatol.* 2006, *vol* 84, 117-126

INDEX

A

abiotic, ix, 85, 144
absorption, viii, 20, 51, 52, 64, 65, 73, 77, 132
absorption spectra, 131, 132
abundance, 151
accounting, 52
accuracy, viii, ix, 12, 19, 54, 67, 71, 74, 76
Acer rubrum, 145
acid, 135, 136
acidic, x, 33, 39, 41, 127, 129, 130, 133, 140, 141
acidification, 128, 139
adaptability, 40
adaptation, 40
adiabatic, 22
adult, 33, 34, 113
adults, 33
aerosol, xi, 153, 154, 155, 158
Africa, 33, 37, 86
afternoon, 157
age, 26, 59, 68, 79, 80, 81, 112, 123, 130, 134, 139
agent, 145
agents, 34, 45
agricultural, ix, x, 38, 85, 87, 103, 106, 108, 117, 118
agroforestry, 151
air, xi, 2, 4, 5, 9, 12, 21, 22, 23, 24, 54, 56, 60, 87,
    90, 91, 93, 113, 118, 128, 133, 134, 137, 139,
    140, 141, 145, 153, 154, 155
air pollutants, 133, 139, 140, 141
Alabama, 38, 45
alcohol, 145
alkalinity, 33
alternative, ix, 37, 71, 150
alters, 31
Amazon, 21, 86, 98, 100
Amazonian, 93, 97
ambient air, 22, 24, 54

ambient air temperature, 54
ammonium, 41
animals, 33, 45, 104, 145
anthropogenic, x, 111, 117, 118, 127, 131, 133, 134,
    141, 157
antimony, x, 127, 134, 140
Appalachian Mountains, 30, 144
application, viii, 13, 17, 25, 34, 37, 45, 48, 52, 68,
    74, 80, 155
Araneae, 150
Arctic, 87, 98
arid, 123
Arizona, 74, 83
arrest, 42
arthropods, 151
ash, 2, 44
ASI, 125
Asia, ix, 20, 35, 37, 53, 85, 86, 87, 99, 101, 139
Asian, 99, 100, 139
assessment, 42, 69, 155
assimilation, 68, 115, 157
assumptions, 67, 77
Atlantic, x, 103, 106
atmosphere, x, 52, 86, 99, 104, 112, 114, 117, 119,
    122, 123, 124, 126, 127, 128, 139, 154
atmospheric deposition, 141
Australia, 28, 37, 99
autotrophic, 87, 89, 113
availability, viii, 37, 52, 60, 69, 74, 80, 124, 144, 151
averaging, 66

B

bacteria, 144, 147
bacterial, 100, 147, 148
banks, 30
bathymetric, 78

behavior, 80, 116, 119, 123, 144
Belgium, 108, 109
benefits, 125
bias, 53
binomial distribution, 146
biocontrol, 45
biodiversity, 40, 45, 48, 114, 144, 152
biogeochemical process, 113
biogeography, 112, 119
biological activity, 93
biological control, 43
biological processes, 113
biomass, viii, ix, x, 30, 36, 39, 40, 71, 72, 78, 80, 82,
    83, 86, 91, 92, 96, 97, 99, 100, 104, 108, 112,
    115, 117, 118, 119, 121, 122, 124
biophysics, x, 111, 113, 118, 119
biosphere, vii, x, xi, 99, 111, 112, 113, 118, 123,
    126, 153, 154
biosynthesis, 131
biota, 144, 152
biotic, ix, 21, 23, 85, 86, 93
biotic factor, ix, 85
bison, 104, 106
blocks, 20
boreal forest, 87, 100, 125, 126
bovine, 107
Brazil, 31, 95, 96, 150
Brazilian, 46, 86
breeding, 44
browsing, 37, 43, 45, 106
buildings, 75, 155
burning, 31, 34, 36, 39, 41, 42, 45
burns, 144

C

Cameroon, 97, 98
Canada, 33, 81, 148
carbon, ix, xi, 52, 53, 60, 68, 69, 70, 72, 80, 85, 86,
    87, 89, 91, 92, 95, 96, 97, 98, 99, 100, 103, 104,
    105, 106, 108, 109, 113, 116, 117, 124, 125, 126,
    144, 152, 153, 154, 156, 157
carbon cycling, 99, 126
carbon dioxide, 68, 100, 156
case study, 125
cation, 140
cattle, x, 98, 103, 106, 107, 108, 109
cavities, 136
cell, 114, 116, 123
CH4, 158
chemical properties, 140
China, 37
chlorophyll, 52

circulation, x, 111, 112
classes, viii, 53, 71, 79, 80, 81, 131, 147
classical, 155
classification, 81
clay, 39
clean air, 140
cleaning, xi, 153, 154
clients, 74
climate change, x, 67, 97, 111, 112, 113, 114, 118,
    119, 124, 125, 126
climate warming, 100, 126
climatic factors, 113
climatology, 126, 157
closure, 11, 64, 136
clouds, 73
clustering, 64
CO_2, ix, x, 52, 54, 56, 61, 68, 85, 86, 90, 94, 95, 96,
    98, 99, 100, 101, 103, 104, 109, 111, 112, 113,
    115, 117, 118, 119, 122, 123, 124, 126, 154, 155,
    157, 158
Coleoptera, 34
collagen, ix, 103, 104, 105, 106, 109
colonization, 113
Colorado, 82, 152
Columbia, 47
combustion, 137, 157
communities, ix, 30, 31, 35, 40, 41, 42, 44, 48, 69,
    71, 76, 78, 150, 151
community, x, 31, 37, 40, 43, 44, 46, 69, 80, 83, 87,
    98, 111, 112, 113, 118, 146, 148, 151, 152, 154
compensation, 40, 54, 60, 70
competition, 30, 31, 40, 45, 112, 113, 114, 118, 147
competitive advantage, 34, 40
competitor, 31, 34
complex interactions, viii, 29
complexity, 59
components, 17, 20, 24, 56, 99, 133
composition, x, 29, 30, 53, 72, 98, 100, 103, 106,
    109, 112, 114, 123, 129, 140, 141, 144, 148, 150
compounds, xi, 128, 137, 139, 158
concentration, xi, 54, 56, 87, 99, 104, 115, 119, 122,
    128, 137, 139, 155, 158
condensation, 24
conductance, 12, 69, 104, 112
conductive, 133
confidence, 150
confidence intervals, 150
configuration, 78
Congress, 27
conifer, 141
coniferous, 27, 53, 140, 141, 150
conservation, 45, 132
constraints, 115

construction, 38, 79
consumption, xi, 106, 128, 138, 139
control, viii, xi, 29, 31, 33, 34, 37, 41, 42, 43, 45, 46,
    47, 48, 68, 79, 143, 144, 146, 147, 149, 152
convection, 22
conversion, 117
Copenhagen, 155, 157
COR, 9
correlation, ix, 5, 6, 27, 52, 58, 76, 85, 91, 93, 96, 97,
    134, 136, 139
correlation coefficient, 5, 6, 139
correlations, 58, 100, 138
Costa Rica, 95, 100
coupling, 37, 114, 119, 123
CRC, 69
credit, 32, 35, 38
croplands, 86
cross-sectional, 78
crowding out, 33
cultivation, 100
cuticle, 128, 140
cyanide, 34
cycles, x, xi, 86, 111, 128, 139, 147
cycling, x, 99, 111, 112, 113, 118, 123, 124, 126,
    140, 144

deposition, x, 127, 128, 134, 139, 140, 141, 154
depressed, 135, 146
depression, 136
desert, 24, 77, 123
detection, viii, 29, 31, 42, 80, 129
deviation, 4, 134
dietary, 104
diffusivity, 87, 90, 93, 96, 98
disability, xi, 127
discrimination, 109
dispersion, 64
displacement, 12, 22, 48
distribution, vii, viii, ix, 5, 21, 26, 32, 33, 36, 39, 44,
    48, 51, 52, 53, 54, 55, 59, 62, 63, 64, 65, 67, 68,
    69, 71, 73, 74, 76, 77, 83, 86, 99, 122, 123, 134,
    140, 146
District of Columbia, 47
divergence, 73
diversity, vii, 34, 36, 40, 44, 47, 151, 152, 155
dominance, 31
Dominican Republic, 48
drought, 31, 37, 40, 48, 49, 125, 147, 148, 150
drying, 14, 16
duration, 3, 5, 6, 11, 12, 70, 73
dust, 134, 140
dusts, 137, 140

D

data analysis, 74
data availability, 80
data collection, 147
data processing, 66
data set, 73, 74, 126
death, 36, 41, 117
decay, 97
deciduous, viii, 44, 45, 51, 53, 59, 68, 69, 70, 77, 78,
    81, 82, 97, 98, 100, 140, 141, 150
decisions, 144
decomposition, 86, 97, 100, 116, 117, 121, 123, 124,
    125, 144, 150, 151
defense, 34, 46, 128
deficiency, 128, 131, 133
deficit, 5, 21, 54, 118
definition, 16, 77
deforestation, 23, 86
degradation, xi, 37, 127, 128, 131, 133, 134, 139
demography, 47, 126
Denmark, 106, 153
density, viii, xi, 9, 12, 30, 33, 34, 36, 47, 52, 54, 58,
    59, 64, 65, 67, 68, 74, 109, 128, 137, 144, 151,
    154, 157
Department of Agriculture, 32, 36, 38, 39, 43, 82
Department of the Interior, 47

E

earth, 116, 118, 123, 124, 125, 126
Earth Science, 111
earthworms, 151
East Asia, 139
ecological, viii, 29, 30, 36, 38, 40, 42, 45, 46, 71,
    112, 113, 114, 119, 123, 124, 126, 144, 148, 151
ecologists, 80, 112
ecology, viii, 29, 43, 70, 75, 112, 113, 118, 148, 151,
    152
ecosystem, vii, 30, 33, 41, 42, 52, 69, 70, 89, 93, 98,
    99, 112, 113, 114, 118, 119, 122, 123, 125, 126,
    143, 144, 147, 148, 151
ecosystems, viii, ix, x, xi, 38, 40, 41, 42, 45, 67, 69,
    71, 72, 85, 86, 87, 88, 89, 91, 93, 95, 96, 97, 98,
    99, 100, 108, 109, 111, 112, 125, 126, 127, 128,
    139, 152, 153, 154
education, 124
electromagnetic fields, 54
electron, xi, 127, 135
elk, 33
e-mail, 103, 127
emission, 72, 154, 156, 157
energy, xi, 21, 27, 73, 74, 79, 112, 117, 118, 124,
    153, 154, 157

England, 33

environment, viii, 42, 52, 53, 55, 56, 59, 60, 62, 64, 65, 66, 67, 83, 87, 104, 109, 133, 140, 158

environmental change, 113, 116

environmental conditions, 60, 104, 109, 112, 114, 116, 123

environmental factors, 56, 57, 59, 86, 87, 90, 130, 141

equilibrium, 22, 24, 119, 120, 122

erosion, 131, 134, 136

estimating, viii, 62, 64, 71, 82, 88, 98, 128

estimators, 82

ethyl alcohol, 145

Europe, 33, 34, 44, 106, 109, 139

evaporation, vii, 1, 2, 3, 4, 5, 11, 12, 13, 14, 15, 17, 19, 20, 21, 23, 24, 25, 26, 27, 143

evapotranspiration, 25, 26, 117, 124, 155

evolution, 44, 100, 109

exchange rate, 154

exclusion, 37, 43, 64

exercise, 65

exploitation, x, 104, 108

exposure, x, 127, 130, 133

extinction, 45, 55

extraction, 145

extrapolation, 67

F

family, 33

farmers, x, 104, 108

fauna, vii, 144, 145, 147

fax, 71

February, xi, 8, 44, 143, 145

feedback, 37, 100, 112, 114, 118, 119, 122, 123, 124, 125, 126

feeding, ix, 103, 104, 106, 144, 148, 151

fern, 30, 45, 46, 48

fertility, 39, 126

fertilization, 72, 124

fiber, 42, 72

fibers, 39

film, 9

filters, 30

Finland, xii, 153

fire, 31, 34, 40, 41, 45, 46, 47, 80, 124, 126, 152

fire suppression, 124

fires, 31, 41, 80

fish, 55

fisheries, 72

flavonoids, 132

flora, vii, 41, 144

flora and fauna, vii

flow, vii, 10, 21, 23, 24, 133, 141, 144, 152

fluctuations, 64

focusing, 3, 144

food, 144, 147

food production, 147

forbs, 34

Ford, 77, 82

forest ecosystem, x, 42, 45, 67, 69, 98, 111, 126, 127, 128, 139, 151, 152

forest fire, 126

forest formations, 109

forest management, viii, 29, 31, 36, 145

forest resources, 72, 104, 107, 108

Forest Service, 43, 47, 82, 145

forestry, 1, 25, 26, 29, 46, 75, 81, 82, 83, 98, 99

forests, viii, ix, x, 25, 27, 29, 31, 33, 34, 35, 39, 40, 42, 44, 46, 47, 48, 68, 70, 71, 77, 78, 81, 82, 83, 85, 86, 87, 91, 93, 96, 97, 98, 100, 101, 104, 109, 112, 118, 119, 121, 125, 127, 139, 141, 147, 150, 157

Fox, 83

fragmentation, 82

France, ix, 29, 95, 103, 104, 106, 108, 109

freezing, 45, 60

frequency distribution, 62

frost, 42, 114

fuel, 137

funding, 150

fungal, 31, 44, 100, 128, 144, 147, 148, 151, 152

fungal infection, 128

fungi, 31, 34, 48, 144, 147, 151, 152

G

gas, 24, 52, 54, 55, 87, 89, 90, 93, 94, 95, 154, 156, 157

gas exchange, 52, 54, 55

gases, x, 127, 129, 130, 133

gauge, 20

generation, 27, 112

genes, 43

geography, 126

Georgia, 38, 152

Germany, 26, 89, 103, 157, 158

germination, 39, 43, 47

glass, 4, 145

Global Positioning System (GPS), 73

global trends, 148

globalization, 45

glyphosate, 34, 42, 45

government, iv

graph, 54

grass, 30, 31, 35, 38, 42, 43, 44, 45, 46, 47, 49, 106

grasses, 30, 31, 44, 45
grassland, 27, 53, 78, 152
grasslands, 39, 86, 126, 157
grazing, 151
greenhouse gas, 157
grid resolution, 118
grids, 112, 118
ground-based, 79
groundwater, 21
grouping, 109
groups, 104, 145
growth, 5, 10, 27, 30, 31, 33, 34, 37, 41, 44, 45, 46, 47, 48, 70, 78, 80, 83, 100, 112, 113, 114, 116, 118, 119, 147, 148
growth rate, 70
guard cell, 135

H

habitat, 47, 49, 72, 76, 80, 104, 106, 108, 148
hardwood forest, 46, 47, 81, 101, 145, 151
hardwoods, 33
harvest, 43, 72
harvesting, 36, 43
Hawaii, 30, 46
haze, 101
heart, 33
heat, vii, 1, 2, 3, 11, 12, 13, 14, 15, 16, 17, 19, 20, 21, 24, 25, 39, 118, 119, 123, 154, 155
heat release, 24
heat transfer, 24, 118
heating, 22, 154
height, ix, 9, 12, 17, 22, 24, 53, 57, 58, 64, 65, 72, 73, 74, 75, 76, 77, 78, 79, 81, 83, 87, 112, 116, 130, 132, 133, 138
herbicide, 34, 37, 42, 47
herbicides, 37, 42, 45, 48
herbivores, 31, 104, 108, 109
herbivorous, ix, 103, 104, 108
herbivory, 30, 34, 37, 38, 46, 48
herbs, 31, 37, 42, 43
heterogeneity, 67, 100, 123, 125
heterogeneous, 155
heterotrophic, ix, 86, 91
high resolution, ix, 72
high temperature, 148
Holocene, x, 103, 106, 119
horizon, 56
host, 45
hot spots, 87
house, 81
human, 33, 107, 117, 119, 124, 129, 137
humans, x, 33, 104, 108

humidity, 2, 3, 5, 9, 12, 19, 24
hunting, 108
hurricane, 144, 145
husbandry, x, 103, 106
hydro, 19, 21, 158
hydrocarbons, 158
hydrologic, vii, 27
hydrological, x, 26, 27, 111, 118
hydrological cycle, 26
hydrology, vii, 1, 2, 9, 27
hypothesis, 40, 43, 44, 46

I

IDEA, 3, 9, 13, 20
identification, 80
Illinois, 37, 43, 45, 47
illumination, 73, 75, 77
images, viii, 71
in situ, 52, 72
independent variable, 78
India, 33, 37
Indiana, 32, 47
indication, 109
indicators, xi, 127, 137
indices, 55, 78, 80
indigenous, 47
indirect effect, 144, 148
Indonesia, 20, 25, 27
industrial, 119, 122, 123
industrialization, 139
inequality, 11
infection, 128
infestations, 37, 40
inherited, 114
inhibition, 30, 31, 34, 46, 47
inorganic, 86, 139, 140, 141
INS, 73
insects, 33, 144
insight, 148
instability, 22
instruments, 9, 73, 74, 158
integration, 52, 80
integrity, 38, 124
interaction, x, 30, 48, 68, 111, 112, 124, 126, 154
interactions, viii, xi, 29, 31, 37, 43, 46, 64, 113, 119, 126, 128, 129, 137, 138, 139, 140, 141, 146, 148, 153, 154
interface, x, 127, 128, 139
interference, 46
Intergovernmental Panel on Climate Change, 125
interrelations, 144, 152
interval, 5, 9, 17, 72

invasive, viii, 29, 30, 31, 34, 38, 41, 42, 43, 44, 45, 46, 47, 48
invasive species, viii, 29, 31, 42, 44, 48
invertebrates, 152
ions, 128, 129, 138
IPCC, 112, 119, 125
Iran, 71
irrigation, 33
island, 129
isotopes, 108, 109
Isotopic, 104
Italy, 33

J

Japan, vii, 1, 2, 3, 7, 9, 12, 25, 26, 27, 37, 38, 47, 51, 53, 55, 69, 70, 85, 89, 98, 100, 111, 127, 128, 133, 137, 140, 141
Japanese, vii, x, 1, 2, 3, 5, 9, 11, 19, 20, 25, 26, 27, 31, 46, 69, 99, 127, 128, 136, 140, 141, 155
Java, 19, 25
Jun, 156

K

K^+, xi, 40, 128, 129, 137, 138, 139
Kentucky, 147
killing, 42
King, 158

L

Lafayette, 29
lamina, 54
land, vii, viii, ix, xi, 26, 29, 38, 49, 69, 71, 83, 85, 86, 87, 99, 100, 112, 113, 114, 117, 118, 119, 122, 123, 124, 125, 144, 153, 154, 157
land use, viii, ix, 29, 49, 85, 86, 99, 100, 117, 118, 119, 124, 125, 144, 157
landscapes, ix, xi, 78, 103, 153, 154
large-scale, 78, 88, 113, 125
laser, 72, 73, 74, 75, 77, 81, 82, 83
laundry, 21
law, viii, 24, 51, 52, 55, 56, 62, 67, 116
LCP, 54, 60
leaching, xi, 128, 129, 138, 139, 140
Leaf area index (LAI), 9
leaf blades, 38
leakage, 20
LED, 54
lens, 55
life forms, 109

light conditions, 31, 55, 62
light transmittance, 81
limitations, viii, 44, 52, 67, 71, 155
Lincoln, 9, 89
linear, xi, 54, 136, 143, 146
linear regression, 54
location, 78, 89
logging, 36, 99, 100, 118, 144, 152
London, 45, 48, 70, 151
long distance, 39
longevity, 147
longleaf pine, 40, 41
losses, 121
Louisiana, 38
low temperatures, 60
low-tech, 152

M

macropores, 27
maintenance, xi, 38, 153
maize, 152
Malaysia, ix, 85, 95, 98, 99, 100, 101
mammal, 109
mammals, ix, 103, 104
management, viii, 29, 31, 34, 36, 41, 42, 48, 72, 97, 99, 118, 124, 145, 152
management practices, 36
manipulation, 144
mapping, viii, 71, 73, 75, 80, 82
Maryland, 47, 82
matrix, 27
measurement, vii, xi, 1, 2, 4, 9, 12, 21, 52, 72, 77, 80, 82, 88, 89, 153, 154
measures, 41, 43, 79, 80
median, 79, 156
Mediterranean, 5, 7, 26
memory, 113
Mercury, 2
meteorological, 5, 9, 12, 21, 27, 118, 146, 154
methylene, 136
Mexico, 147, 148, 151, 155, 157
Mexico City, 155, 157
microbes, 93, 144
microbial, 31, 35, 46, 87, 98, 117, 121, 150
microbial communities, 31, 35
microbial community, 31, 46, 87, 98
microclimate, 154
micrometeorological, 2, 20, 26, 154, 155
microorganisms, ix, 85
microscope, 145
migration, 147
mineralization, 150

Ministry of Education, 124
mirror, 73
misconception, 73
missions, 99, 157
Mississippi, 38, 40, 49
mites, xi, 143, 144, 145, 146, 147, 150
mixing, 24
model system, 119, 123
modeling, viii, 16, 42, 52, 65, 71, 78, 80, 119, 124
models, vii, viii, x, 1, 2, 3, 9, 11, 13, 14, 15, 16, 18, 19, 25, 26, 27, 45, 51, 52, 53, 55, 56, 62, 63, 64, 65, 66, 67, 68, 69, 70, 87, 91, 92, 96, 111, 112, 118, 119, 124, 125, 126, 152
moisture, 21, 23, 26, 37, 115, 116, 117, 118, 123, 124, 125, 143, 146, 147, 148, 150, 152
molecules, 43
Møller, 155, 157, 158
momentum, 12, 154, 155
monsoon, 24
morphological, 35, 70
mortality, 34, 41, 112, 113, 114, 118, 119
motion, 22
movement, 42, 119, 121, 148, 155
multiple regression, 79, 136
multiple regression analysis, 136
mycelium, 144

N

nanometers, 73
NASA, 74
National Science Foundation, 150, 152
native plant, 33, 34, 35, 37, 40, 41, 48
native species, 31, 36, 37, 40, 41, 43
NATO, 125
natural, xi, 9, 33, 36, 39, 40, 42, 43, 44, 47, 74, 77, 87, 109, 114, 123, 130, 131, 141, 143, 144, 145, 150, 153, 154, 157
natural environment, 130, 141
natural gas, 154, 157
needles, 129, 140, 141
negative relation, 90
nematode, 151
nematodes, 144
Nepal, 31, 47
nesting, 80, 82
Netherlands, 83
network, xi, 39, 40, 153, 154
New York, 26, 30, 33, 45, 69, 125, 126, 140, 141, 151
New Zealand, 30
Newton, 151
Nigeria, 147

nitrogen, 44, 52, 68, 69, 109, 115, 139, 140, 141, 144, 150, 151, 152, 158
nitrogen oxides, 158
nitrous oxide, 101, 158
non-destructive, 68
non-linearity, 64
non-native, 45
non-randomness, 52
North Africa, 33
North America, 31, 34, 35, 45, 47, 48
North Carolina, xi, 43, 143, 145
Norway, 78, 141, 147
Norway spruce, 78, 141
NPP, 122, 123
nutrient, 33, 97, 124, 140, 144
nutrient cycling, 140
nutrients, 30, 40, 144
nutrition, 46, 68

O

oat, 30
obligate, 33
observations, 27
occlusion, 77
Ohio, 45, 47
oil, ix, 31, 85, 86, 87, 88, 89, 90, 91, 92, 93, 96, 97, 98, 141, 143
oil palm, 94, 95, 97
oils, 33, 39, 145
one dimension, 116
online, 126
openness, 45
optical, viii, ix, xi, 71, 72, 127
optical systems, viii, 71
optimization, 19
orbit, 60
Oregon, 83
ores, 147
organic, ix, 39, 85, 86, 97, 98, 99, 109, 112, 113, 117, 124, 125, 126, 152
organic matter, 39, 97, 98, 112, 125, 152
orientation, viii, 51, 52, 54, 64, 67, 68, 73, 77
oversight, 15
oxidants, 128, 134
oxide, 101, 158
ozone, 139

P

Pacific, 9, 37, 81, 82
Panama, 99

parameter, vii, 1
Paris, x, 104, 107, 108
particles, xi, 127, 153, 154, 158
particulate matter, x, 127, 128, 129, 133, 134, 135, 136, 137, 139, 140
partition, 151
passive, viii, ix, 71, 72
pasture, 96, 118
pastures, 98, 108
pathways, 33
peat, 125
Pennsylvania, 48
permafrost, 124
pesticide, 144
pesticides, 152
pests, 40
pH, 39, 41, 88
phosphate, 48
photographs, 55, 56, 59, 64
photon, 54, 68
photons, 61
photosynthesis, vii, viii, 47, 51, 52, 53, 54, 55, 56, 59, 62, 63, 64, 65, 66, 67, 68, 69, 70, 86, 101, 112, 114, 115
photosynthetic, 33, 47, 52, 54, 59, 60, 61, 62, 63, 68, 69, 82, 104, 114, 123
physical properties, 22
physics, 91, 99, 123
physiological, viii, 52, 53, 68, 70, 112, 124, 128, 133, 141
physiology, xi, 48, 70, 128, 139
pigments, 52
pitch, 73
plants, viii, ix, 29, 30, 31, 32, 33, 34, 35, 36, 37, 38, 39, 40, 41, 42, 43, 44, 47, 48, 60, 68, 69, 70, 100, 103, 104, 105, 106, 109, 115, 139
plastic, 10, 35, 145
plasticity, 33, 40, 44
play, 53
Poland, 159
pollination, 33
pollutants, 128, 133, 139, 140, 141
pollution, xi, 128, 134, 137, 139, 141
polygons, 54
polymer, 9, 12
polyurethane, 10
pools, 86, 117, 119, 120
poor, 33, 40
population, x, xi, 30, 34, 37, 47, 108, 109, 111, 112, 113, 114, 118, 128, 137, 144, 153
population density, xi, 47, 109, 128, 137, 144
population growth, 30, 34
positive correlation, ix, 85, 96, 97, 134

positive feedback, 100, 112
positive relationship, 92, 147
power, 73, 74, 77, 78
precipitation, 8, 9, 16, 21, 22, 23, 53, 87, 92, 112, 113, 118, 136, 141, 145, 150
predators, 108
predictability, 123, 124
prediction, 78
predictive model, 27
predictor variables, 146
pressure, viii, 5, 12, 21, 22, 23, 24, 29, 54, 56, 107, 118
prevention, 42
private, 42, 72, 74
probability, 77, 88, 116
probe, 89
production, x, 30, 33, 34, 37, 40, 42, 69, 72, 86, 93, 98, 99, 100, 111, 122, 125, 132, 133, 134, 147
productivity, ix, 5, 33, 41, 44, 70, 71, 86, 123
program, 42, 55, 68
propagation, 124
protection, 98
public, 42, 72, 74
Puerto Rico, 5, 6, 7, 19, 27, 100
pulse, 72, 73, 74

Q

quantum, 54, 55, 60, 69, 83
Quercus, 46, 53, 145

R

radar, ix, 72, 82
radiation, x, 9, 12, 21, 26, 53, 62, 68, 69, 77, 112, 127, 132, 155
radius, 3
rain, x, 2, 3, 5, 6, 7, 9, 10, 11, 13, 14, 15, 16, 17, 18, 19, 20, 21, 25, 27, 86, 87, 99, 100, 101, 127, 133, 139, 147, 151
rain forest, 19, 25, 27, 86, 87, 99, 100, 101, 147, 151
rainfall, vii, 1, 2, 3, 4, 5, 6, 7, 9, 10, 11, 12, 13, 14, 15, 16, 17, 19, 20, 21, 24, 25, 26, 27, 38, 92, 129
rainforest, 5, 6, 25, 26, 28, 99, 100, 151
rainwater, 11, 14, 19
random, 146
randomness, 52
range, 2, 30, 33, 48, 54, 73, 106, 115, 148, 157
rangeland, 83
rape, 104
recovery, 43, 49, 119, 124, 144, 147, 152
recreation, 72

recruiting, 37
recycling, ix, 103, 104, 152
redundancy, 114
reflection, 72
refuge, 106
regeneration, 30, 31, 33, 40, 43, 45, 46, 48, 87, 152
regional, x, 42, 67, 101, 104, 108, 112, 123, 146, 148
regression, 3, 4, 5, 6, 7, 10, 13, 14, 15, 54, 92, 134, 136, 157
regression analysis, 136
regression equation, 92
regression line, 3, 4, 5, 6, 7, 10, 13, 14, 134
regressions, 79
regrowth, 39, 42, 147
regular, 139
relationship, 40, 52, 54, 87, 91, 92, 93, 100, 125, 136, 137, 139, 147, 148
relationships, 78, 92, 93, 98, 113, 147, 148
remote sensing, viii, ix, 52, 71, 74, 79, 81, 82
reproduction, 37, 39, 46, 47
residential, 157
residues, 46
resilience, 37
resistance, 12, 40, 79, 113, 133
resolution, ix, 10, 16, 72, 73, 81, 118, 158
resource allocation, 46
resources, 30, 31, 40, 72, 104, 107, 108
respiration, ix, 54, 60, 85, 86, 87, 88, 89, 90, 91, 92, 93, 94, 95, 96, 97, 98, 99, 100, 101, 112, 113, 115, 122, 123, 157
restructuring, 145
retention, 73
returns, 73, 76, 77, 144
Reynolds, 26, 143, 150
rhizome, 38, 39, 40, 42
rhizosphere, 100
rice, 100
risk assessment, 42
rivers, 33
robustness, 123
rods, 77
roughness, 12, 20, 78, 82, 83, 154, 155
routines, 155
Royal Society, 126, 151
rubber, ix, 85, 86, 87, 88, 89, 90, 91, 92, 93, 94, 95
runoff, 9, 126

S

salt, 138
sample, 58, 59, 86, 88, 98
sampling, xi, 77, 78, 83, 88, 128, 133, 137
sand, 39

SAR, 72
SAS, 123, 146, 152
satellite, 42, 74
saturation, 14, 16, 21, 62, 116
scaling, 68, 77, 126
Scandinavia, 109
scatter, 5, 6, 13
scattered light, 64
Schmid, 155, 158
Scots pine, 78, 82, 140
sea ice, 122
sea level, 145
sea-salt, 138
seasonal variations, 5
seasonality, 112
seed, 30, 33, 34, 35, 37, 39, 40, 42, 113, 118, 119
seedlings, 30, 31, 34, 37, 40, 41, 45, 46, 48, 69, 87, 151
seeds, 33, 34, 39
segregation, 48
selecting, 67
selectivity, 144
SEM, 129, 140
semi-natural, 151
sensing, viii, ix, 52, 71, 75, 80, 81, 82
sensitivity, viii, 19, 41, 71, 99, 116, 119, 125, 133
sensors, viii, ix, 55, 69, 71, 72, 73, 75, 77, 79
separation, 3, 9, 13
series, 15, 41, 74, 77
services, iv, 42
severity, 37
sexual reproduction, 39
shade, 31, 34, 35, 40, 43, 65, 70
shape, 54
shock, 119
shoot, 39, 70
short period, 64, 144
short-term, 112, 152
signs, 133
silica, 38
silicates, 38
simulation, 3, 26, 53, 56, 62, 67, 69, 113, 116, 119, 120, 122, 123, 124, 125
simulations, vii, x, 53, 64, 66, 111, 118, 119, 120, 124
sine, viii, 10, 13, 51, 55, 62, 63, 64, 67
Singapore, 81
sites, vii, ix, x, xi, 1, 2, 5, 7, 9, 16, 20, 25, 33, 35, 40, 42, 43, 85, 89, 95, 97, 99, 103, 106, 107, 108, 109, 133, 143, 144, 145, 146, 147, 148, 153, 154, 155, 157
skills, 72
SLA, 115

SO2, 133
SOC, ix, 56, 85, 86, 87, 91, 92, 96, 97, 113, 116,
    117, 118, 119, 121, 122, 123, 124
soil, vii, ix, x, xi, 27, 31, 35, 38, 39, 40, 41, 46, 48,
    85, 86, 87, 88, 89, 90, 91, 92, 93, 94, 95, 96, 97,
    98, 99, 100, 101, 109, 111, 112, 113, 115, 116,
    117, 118, 121, 123, 125, 126, 128, 140, 143, 144,
    146, 147, 148, 149, 150, 151, 152, 157
soils, 33, 39, 40, 87, 101, 152
solar, 2, 21, 26, 53, 62, 69, 112
solid phase, 91
sounds, 2
South America, 30
South Carolina, 38
Southeast Asia, ix, 85, 87, 101
Spain, 5, 7
spatial, ix, 27, 30, 42, 44, 45, 52, 54, 55, 59, 65, 67,
    71, 73, 74, 78, 80, 85, 96, 99, 100, 101, 109, 150
spatial analysis, ix, 71
spatial heterogeneity, 67
spatiotemporal, 123
species, viii, ix, 5, 29, 30, 31, 32, 33, 34, 36, 37, 39,
    40, 41, 42, 43, 44, 45, 46, 47, 48, 49, 53, 60, 68,
    69, 71, 72, 79, 80, 81, 98, 100, 104, 108, 112,
    113, 114, 118, 131, 132, 140, 144, 147, 148, 150,
    151
species richness, 30, 40
specific heat, 12
specificity, 37
spectrum, 144
speed, 9, 12, 21, 72, 78, 118, 154, 155
speed of light, 72
Sri Lanka, 33
stability, 12, 22, 40, 42, 44
stabilization, 119
stabilize, 38
stages, 14
standard deviation, 4, 130, 131, 132, 134
statistical inference, 152
stochastic, 121, 123, 124
stock, x, 111, 116, 118
storage, x, 9, 11, 19, 20, 26, 74, 86, 97, 99, 111, 112,
    113, 117, 123
storms, 14, 17, 20
strategies, 40
stratification, 150, 154
streams, 35
strength, 76
stress, x, 113, 127, 128, 133, 136, 139
stress factors, 113
structural characteristics, 81
suburban, 157
sugar, 69

sulfur, 140
summer, 5, 9, 13, 34, 130, 155, 157
sunlight, 133, 143
supply, 25, 86, 91, 97
suppression, 34, 46, 124
surface area, vii, 1, 3, 4, 131, 132, 134
surface energy, 118, 124
surface layer, 52, 154
surface properties, xi, 128, 133, 137, 138, 139
surface roughness, 154
surface water, 112
surprise, 125
survival, 40, 45, 47, 113, 116, 118
surviving, 60
swamps, 39
Sweden, 33
symbols, 66, 136
systems, 33, 73, 74, 78, 144, 150, 154

T

taxa, 144, 146, 147, 148
technology, ix, 48, 71, 152
teeth, x, 103, 106
temperature, xi, 5, 9, 12, 21, 23, 34, 53, 54, 56, 60,
    86, 87, 88, 89, 90, 92, 93, 96, 97, 98, 113, 115,
    116, 118, 119, 121, 122, 123, 124, 125, 136, 143,
    145, 146, 147, 148, 149, 153, 154, 155
temporal, ix, xi, 48, 52, 59, 65, 66, 69, 74, 85, 87, 96,
    99, 100, 101, 109, 153
temporal distribution, 74
Tennessee, 44, 47
Texas, 38, 89
Thailand, 99
thinking, 21
threat, 41
threatening, 38
three-dimensional, ix, 69, 71, 74, 75, 76, 78, 116
threshold, 3, 14, 72
thresholds, 139
timber, 43, 72, 75, 78, 82, 83
time consuming, 67
time lags, 118
time resolution, 16, 158
timing, 120
tissue, 115, 133
Tokyo, x, 69, 89, 101, 127, 128, 133, 136, 139, 157
tolerance, 40, 148
tomato, 41
topographic, ix, 72, 74, 75, 80, 82
topsoil, 88
torus, 51
toxic, 34

tracers, x, 103, 106
tracking, x, 104, 108, 118
traffic, 118, 155, 157
traits, 115, 144
trajectory, 113
transfer, vii, 1, 12, 21, 24, 67, 118, 154
transition, 30, 43
transitions, 125
transpiration, xi, 69, 127, 134, 136, 154
transport, 2, 21, 24, 25, 27, 155
transportation, 38, 137
trees, x, xi, 10, 20, 26, 30, 42, 43, 53, 72, 79, 80, 81,
    82, 87, 113, 116, 117, 127, 128, 132, 133, 136,
    140, 143, 145, 147, 151
triggers, 22
tropical areas, 20
tropical forest, ix, 68, 70, 81, 83, 85, 86, 92, 93, 96,
    97, 98, 99, 100, 101, 104, 109, 150
tropical rain forests, 86, 87
tundra, 112, 113, 118, 121
turbulence, 154, 155
turbulent, 154
turnover, ix, 49, 86, 97, 99, 116, 119, 125, 147
two-dimensional, ix, 71
two-way, x, 111, 112, 118, 119, 122

U

U.S. Department of Agriculture, 43
ultraviolet, 140
uncertainty, 97
United States, 33, 35, 37, 38, 43, 44, 45, 139
urban areas, xi, 33, 75, 153, 155
USDA, 47, 71, 145
USSR, 33
UV absorption spectra, 131, 132
UV radiation, x, 127, 132

V

validation, 67, 69, 82, 125
values, 4, 5, 7, 8, 9, 10, 13, 14, 16, 17, 19, 20, 66, 73,
    104, 106, 107, 108, 134, 136, 146, 148, 155, 156,
    157
vapor, vii, 1, 2, 5, 21, 22, 23, 24, 25, 54, 56, 112, 118
variability, 64, 69, 100, 119, 124
variables, 6, 15, 19, 21, 78, 113, 118, 122, 123, 136,
    146, 158
variance, 78
variation, 9, 44, 47, 49, 69, 86, 87, 91, 92, 93, 96, 99,
    100, 101, 141

vegetation, vii, viii, ix, x, xi, 9, 30, 33, 34, 40, 41, 43,
    51, 52, 53, 60, 65, 71, 72, 73, 75, 76, 77, 79, 81,
    82, 83, 86, 97, 98, 99, 100, 103, 106, 111, 112,
    113, 114, 117, 118, 119, 121, 122, 123, 124, 126,
    153, 154, 155, 157, 158
vein, 38
ventilation, 93
Vermont, 69
vertebrates, 34
visible, 140

W

water, vii, ix, x, 1, 2, 4, 5, 9, 10, 11, 19, 20, 21, 22,
    23, 24, 25, 40, 56, 68, 72, 73, 79, 81, 85, 86, 88,
    89, 90, 91, 92, 93, 94, 97, 98, 112, 118, 119, 124,
    125, 127, 128, 129, 130, 131, 133, 135, 136, 138,
    139, 140, 141
water quality, 72
water vapor, vii, 1, 2, 5, 21, 22, 23, 24, 25, 56, 68,
    112
watershed, 2, 83, 145, 147, 151
waterways, 33
wavelengths, 73
weapons, 44
weathering, 140
web, 144
weedy, 43
well-being, xi, 153
Western Europe, 106
wetlands, 157
wettability, xi, 128, 137, 138, 139
wetting, 11, 14, 16, 17, 20
wheat, 104
wild animals, 104
wildlife, ix, 71, 72, 76
wind, 9, 12, 21, 24, 26, 40, 78, 118, 133, 138, 145,
    154, 155
winter, 9, 13, 33, 36, 47, 96, 128, 137, 138, 148, 157
wood, 30, 72, 125
woodland, 44, 48, 78
woods, 33, 109

X

xylem, 141

Y

yield, 54, 60, 73, 77

DATE DUE

DUE DATE SUBJECT TO CHANGE
IF A RECALL IS REQUESTED